Turtles

OF NORTH AMERICA

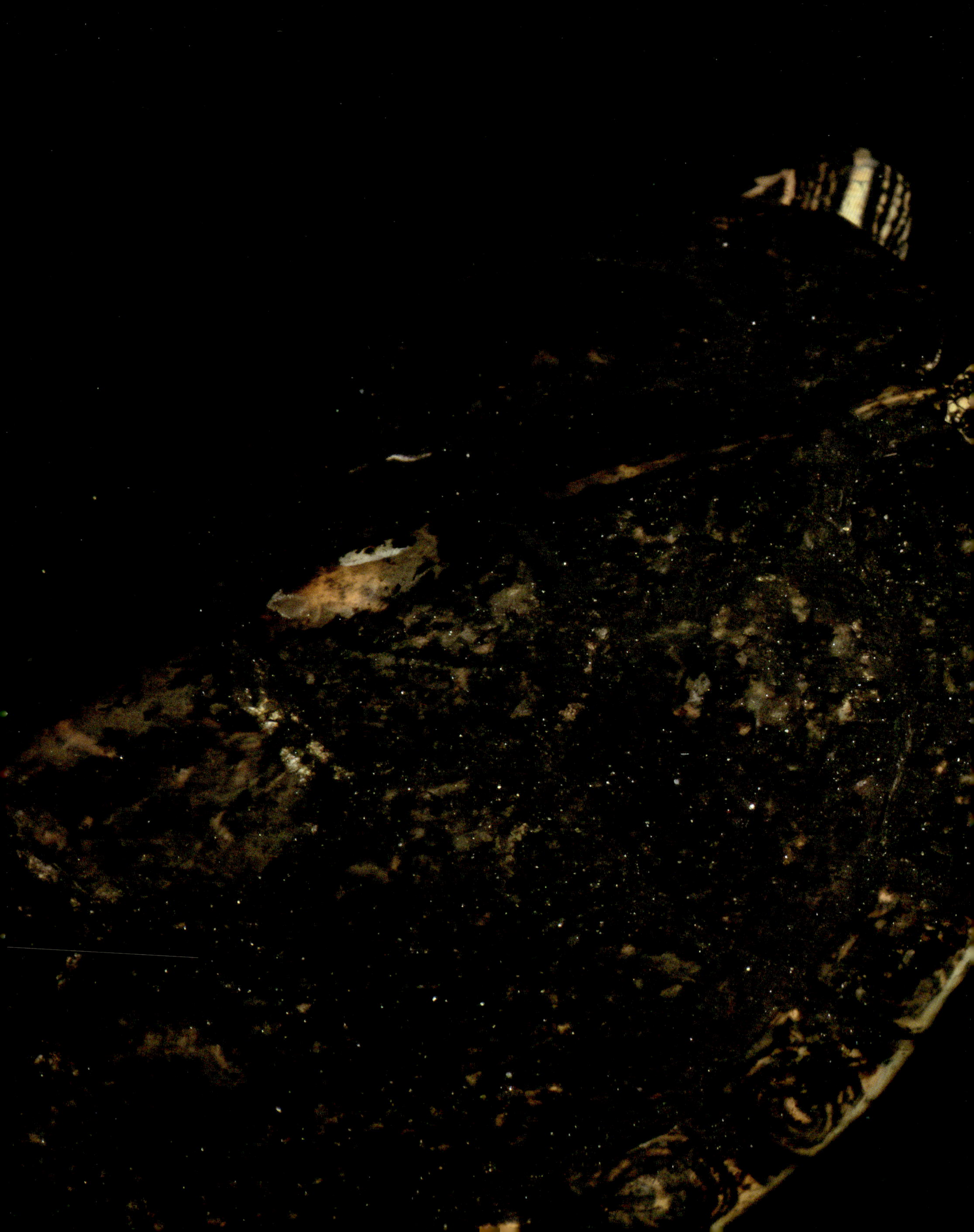

Turtles

OF NORTH AMERICA

An Illustrated Field Guide to the Turtles of the
Continental United States and Canada

KYLE HORNER

FIREFLY BOOKS

A FIREFLY BOOK

Published by Firefly Books Ltd. 2024
Copyright © 2024 Firefly Books Ltd.
Text copyright © 2024 Kyle Horner
Photographs © as listed on page 205 or next to
   individual photographs

All rights reserved. No part of this publication may be
reproduced, stored in a retrieval system, or transmitted in any
form or by any means, electronic, mechanical, photocopying,
recording or otherwise, without the prior written permission
of the Publisher.

First printing

Library of Congress Control Number: 2023946763

Library and Archives Canada Cataloguing in Publication
Title: Turtles of North America : an illustrated field guide to
   the turtles of the continental United States and Canada /
   Kyle Horner.
Names: Horner, Kyle, author.
Description: Includes index.
Identifiers: Canadiana 20230542654 |
   ISBN 9780228104667 (softcover)
Subjects: LCSH: Turtles—United States—Identification. |
   LCSH: Turtles—Canada—Identification. |
   LCSH: Turtles—United States. | LCSH: Turtles—Canada. |
   LCGFT: Field guides.
Classification: LCC QL666.C5 H66 2024 |
   DDC 597.92097—dc23

Published in the United States by
Firefly Books (U.S.) Inc.
P.O. Box 1338, Ellicott Station
Buffalo, New York 14205

Published in Canada by
Firefly Books Ltd.
50 Staples Avenue, Unit 1
Richmond Hill, Ontario L4B 0A7

Cover and interior design: Stacey Cho
Illustrations: George A. Walker
Maps: Kyle Horner

Printed in China

Canada

We acknowledge the financial support
of the Government of Canada.

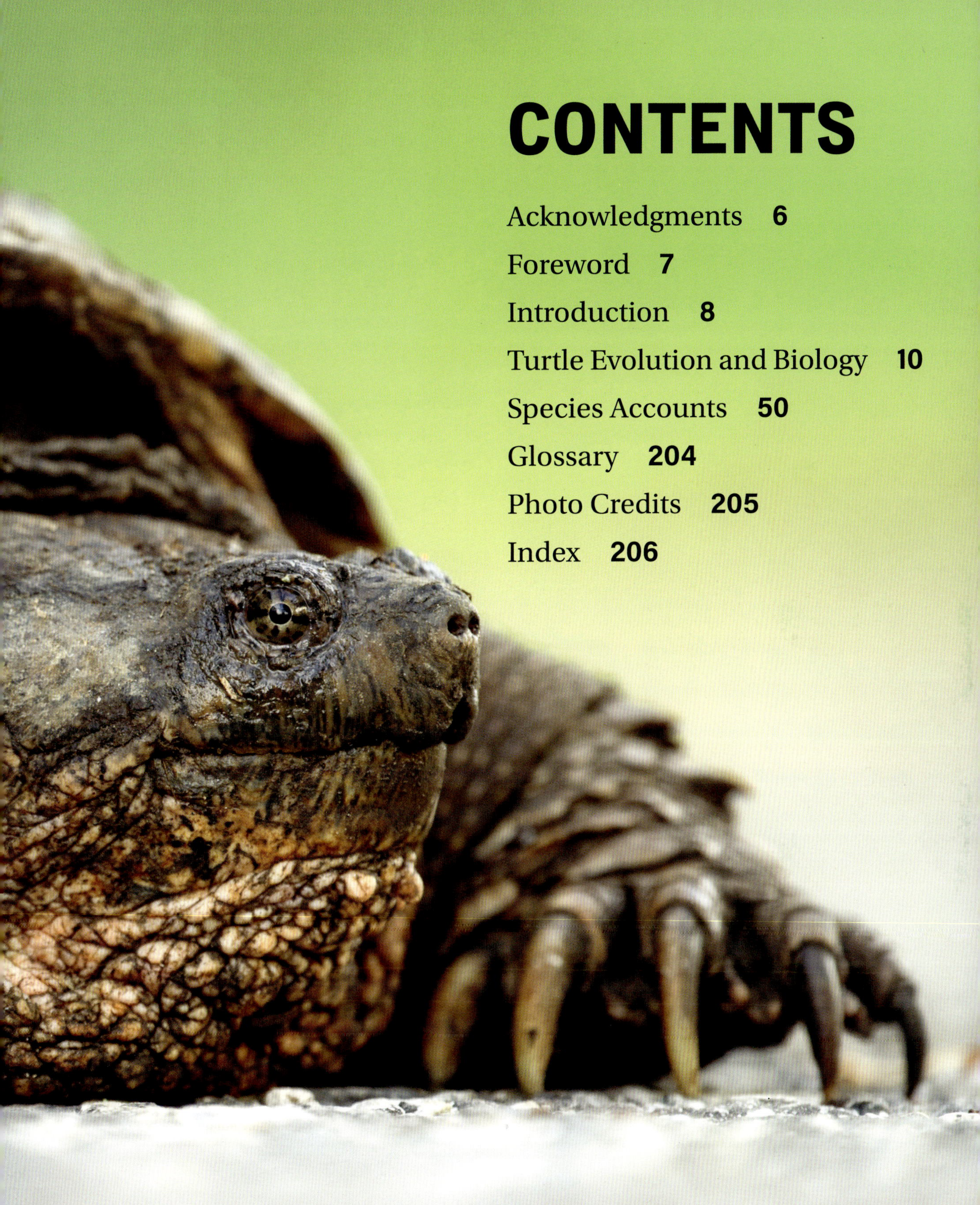

CONTENTS

ACKNOWLEDGMENTS

Thank you to all the naturalists and photographers who provided images for this book. Photographing turtles in the field — especially rare and reclusive species — can be a major challenge, and this book would not be possible without your generous and beautiful contributions. Additional thanks to Grover Brown, Evan Grimes, Kevin Hutcheson, and Carl Franklin and Viviana Ricardez at Texas Turtles for providing multiple images, and dealing so patiently with my many requests and additions.

Thank you to Stacey Cho for your beautiful work in designing this book and your endless accommodations through the process. The finished product looks better than I could have imagined. Thank you to George Walker for your excellent illustrations, and Michael Worek for your guidance.

Finally thank you to Jenny Pearce for facilitating some hard-to-find photos, and for teaching me everything I know about reptile education. I heard your voice in my head several times while writing, and if the occasional sentence in this book sounds familiar to you, that is undoubtedly your influence showing through.

FOREWORD

Turtles are among the most threatened groups of vertebrates on the planet. Of the hundreds of species worldwide, well over half are in danger of becoming extinct due to habitat loss, road mortality, disease, commercial fishing, harvesting for food or sale as pets. However, there is hope — many individuals and organizations are working hard to ensure that no further extinctions occur.

Global organizations such as Turtle Survival Alliance, national organizations such as Ontario Turtle Conservation Centre, as well as many local organizations, are all putting their skills together to protect turtles.

Turtle conservation is appealing to almost everyone — and everyone has their own motivation. Yes, turtles are vital to the biodiversity and health of their environments, and these environments are also vital to human health. However, turtles are also so very cool; they have been around longer than the dinosaurs, they have physiological abilities not found in any other species, and we still have a lot to learn from and about these fascinating creatures. What other species can survive underwater during freezing winters, or can store sperm for years? They also have an amazing ability to heal from injuries that would kill any other species.

Turtles admitted to our hospital at Ontario Turtle Conservation Centre, are brought from across the province and beyond, with life-threatening injuries, mainly due to being hit by cars while crossing the road. Amazingly, the majority of these turtles are healed and later released back to their home wetlands. In addition, their eggs that would have been lost, are incubated at the Centre and the hatchlings are also released back to their mother's home wetland.

Everyone can play a part in turtle conservation; from individual stewardship actions to larger community projects. These Citizen Science activities can play a huge part in adding to the conservation initiatives being done by research biologists across North America and beyond.

Sue Carstairs, B.Sc., D.V.M., Ont
Executive and Medical Director
Ontario Turtle Conservation Centre
https://ontarioturtle.ca

INTRODUCTION

There may be no better-known or more beloved group of animals in North America than turtles. These slow-moving, shell-wearing, swamp-loving reptiles seem to swim into our hearts at an early age.

My own fondness for these animals probably started in my childhood obsession with the Teenage Mutant Ninja Turtles. I spent countless hours watching and re-watching the adventures of Leonardo, Michelangelo, Donatello, and Raphael as they fought crime on the streets of New York. Whether it was their good deeds, love of pizza, or affable nature, something about these "heroes in a half-shell" really struck a chord with my younger self.

Turtles are, in retrospect, an odd choice for superheroes, and you probably had to be the right age at the right time to appreciate them.

But whenever and wherever you grew up, I am certain there were fictional turtles for you to connect with. Whether it was Yertle, Franklin, Crush, Squirtle, the Koopa Troopas, or some other adaptation, turtles were a part of almost everyone's formative years.

Turtles are also common in mythology and folklore both modern and very old. The ancient Greek storyteller Aesop gave us the well-known fable "The Tortoise and the Hare," in which the tortoise is a model of honest perseverance. In cultures worldwide, turtles symbolize longevity, protection, wisdom, and even fertility. The stories of many North American Indigenous groups tell of the world being created and carried upon the back of a turtle, and many such groups refer to the North American continent as Turtle Island.

With all the favorable representations and positive portrayals, it's not difficult to see why children and adults alike love turtles. It is only natural that after appreciating turtles in books and movies, we want to observe and identify them in nature, too. That is where this guide comes in.

This book begins with an introduction to turtle biology, starting with an overview of how turtles came to be, how they diversified, and how we classify the groups of turtles that exist today. A section about anatomy details the basic turtle body plan, and how it varies to take advantage of different environments. A section about behavior covers how they live, from what they eat to what eats them, and ultimately how they make more turtles. Finally, a section about turtle conservation describes some of the common threats to turtles, and how we can take positive action to help them survive and thrive.

Following this introduction to turtle biology, you will find the Species Accounts. These pages describe the 64 species of turtles that inhabit Canada, the continental United States, and the ocean waters that surround them. They show how to identify each species, include a range map that indicates where it may be found, and explain a little about its life history. Numerous insets highlight interesting and unusual aspects of the lives of these turtles.

It is my hope that this book helps you deepen your appreciation for turtles, learn a little about them, and maybe share that love and knowledge with those around you.

THE TORTOISE AND THE HARE

TURTLE EVOLUTION AND BIOLOGY

Proganochelys

TURTLE EVOLUTION

The first turtles lived hundreds of millions of years ago in in the Permian Period. Dinosaurs had not yet walked the earth, but both the land and sea were full of life. The oceans were inhabited by invertebrates like corals, sponges and shellfish, as well as a great array of fish. Amphibians and early reptiles lived in the water too, but many members of both groups had also made the transition to life on land.

Insects were already abundant, including several groups that we would recognize such as early dragonflies and beetles. We might recognize the amphibians, too, though many were much larger than the frogs and salamanders of today. Some looked more like dinosaurs, or even modern crocodiles (bottom of opposite page).

Perhaps the most surprising group of animals in this period were the reptiles, as many looked very different than the modern reptiles we know. The most prominent reptiles in the Permian belonged to a group called the Therapsids, the ancestors of modern mammals. Indeed many would have looked very mammalian to us, complete with fur or hair. Some of these ancestral mammals were plant-eaters, but others — like the incredible Gorgonopsians — were ferocious hunters.

Gorgonops, a Permian Gorgonopsian

The presence of these and other fearsome predators must have influenced the evolution of the other reptiles that lived during the Permian Period. One such group of reptiles would eventually evolve into the dinosaurs and, ultimately, modern birds. Another group would evolve into snakes and lizards and another still into crocodiles and alligators. A final group — and the one we are most interested in for the purposes of this book — was on its way to becoming turtles.

If we could see the early ancestors of turtles, we might easily mistake them for lizards. Like many early reptiles, they had a lizard-like body-plan: a head in line with the body, four legs, and a long tail. Just like modern lizards, these vulnerable creatures would have been popular prey for the larger, carnivorous animals that lived around them.

This perilous life, and the resulting need for protection, may have been what pushed these lizard-like animals to evolve in the direction of modern turtles. We can see the earliest stages of turtle evolution in

Eunotosaurus

Platyoposaurus, a crocodile-like Permian amphibian

a reptile called *Eunotosaurus*. This creature lived around 260 million years ago, and its fossils show us an animal which looks like a lizard at first, but with an unusually broad, rounded ribcage. Each rib is flattened and overlapping the next like a shingle, almost giving the appearance of a shell under the skin.

The unusual ribcage of *Eunotosaurus* may have helped protect it from predators, but not in the same way that shells protect modern turtles. Its "shell" was inside its body and would have done nothing to guard its exposed head and neck. Instead of acting as armor, it seems that this stiff, short torso might have made *Eunotosaurus* a powerful digger, allowing it to burrow underground to escape the dangers on the surface.

During the following tens of millions of years, relatives of *Eunotosaurus* grew gradually closer in appearance to the turtles of today. The first fossil of an animal that really looks like a modern turtle comes from about 210 million years ago, roughly 20 million years after the first appearance of dinosaurs. This creature, called *Proganochelys*, had a complete upper and lower shell that were big enough to be used as protective armor against predators.

In the 210 million years between *Proganochelys* and today, turtles diversified and occupied almost every habitat on the planet. Today they swim in lakes, rivers, ponds, marshes and swamps in every corner of the globe, but they also plod through forests, prairies and even deserts. They have colonized the oceans, with sea turtles adapting the tools of their land-dwelling ancestors to a fully marine lifestyle.

From musk turtles that would fit in the palm of a hand to the cow-sized Leatherback, modern turtles come in all sizes. Their shape, though, is much more consistent. There are some variations, but the basic body-plan of a turtle is the same all over the world. It has barely changed over hundreds of millions of years that included the rise and fall of the dinosaurs, at least two mass extinction events and an ice age. Their ancient form simply works, and it seems there is little need for improvement.

The next time you see a turtle in your local pond or wetland, take a moment to imagine yourself a hundred million years in the past. All of the plants and animals around you, even the atmosphere and the position of the continents on our planet, would be vastly different from today. The turtle in front of you, though, would probably look very much the same.

TURTLE TAXONOMY

Taxonomy is the way we sort and classify living things according to their relationships with other living things. We know that turtles are reptiles, for example, as they share their scaly skin with all the other members of that group. The diagram below shows the most likely relationship of turtles to other modern reptiles, including birds which are the present-day descendants of the prehistoric dinosaurs.

There are approximately 360 species of turtles in the world, and they live on every continent except Antarctica. To make sense of this big group of animals, scientists divide it into smaller groups using the turtles' physical characteristics. The first big division sorts the turtles into two groups based on how each turtle's neck works.

The side-necked turtles (called the Pleurodira) protect their heads by simply bending their necks

REPTILE CLADOGRAM

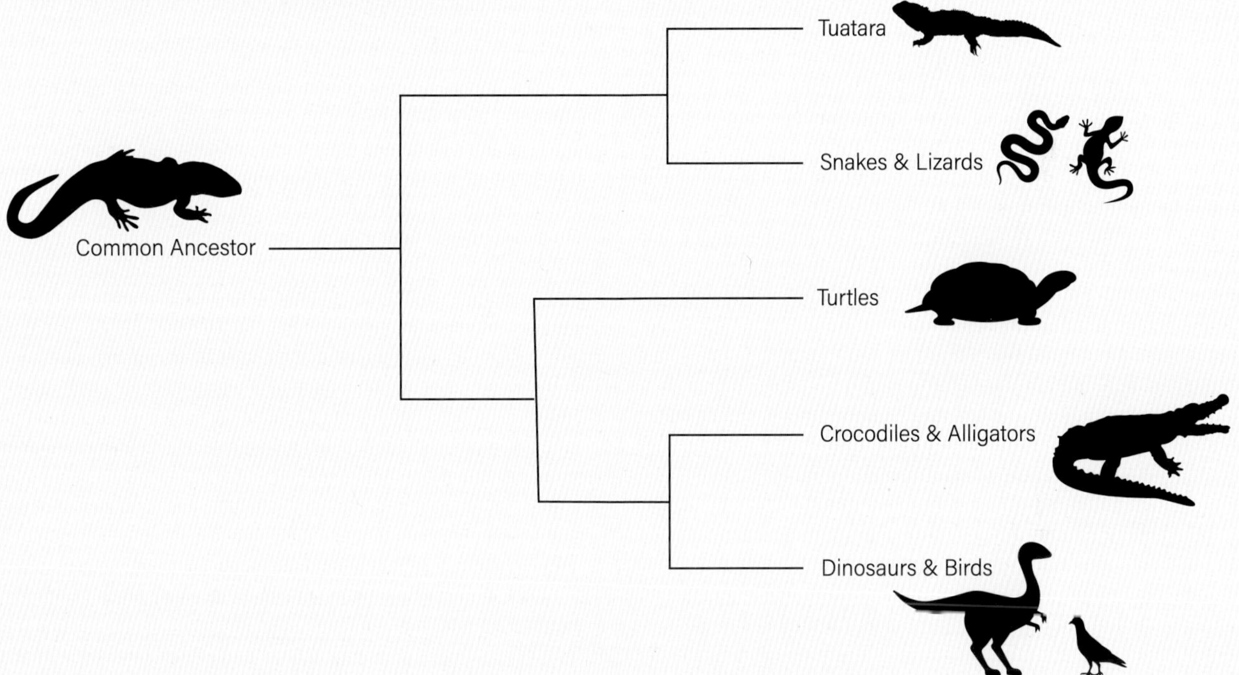

This diagram, called a cladogram, uses lines to show how closely different groups of animals are related to each other. All of these groups are reptiles. Our best current evidence suggests that the turtles are most closely related to crocodilians, dinosaurs, and modern birds, but are separated from these relatives by hundreds of millions of years of evolution.

The African Helmeted Turtle (**LEFT**) and North American Painted Turtle (**RIGHT**) show the two different strategies of neck retraction.

sideways until the entire head and neck is hidden under the front edge of the shell. Approximately 80 species of modern turtles move their necks this way, and all of those live in the southern hemisphere, in places like South America, Africa and Australia.

The hidden-necked turtles (called the Cryptodira) are able to bend their necks vertically into an "S" shape and draw their heads straight back into their shells between their front legs. This group includes all the remaining turtle species in the world, including every turtle native to North

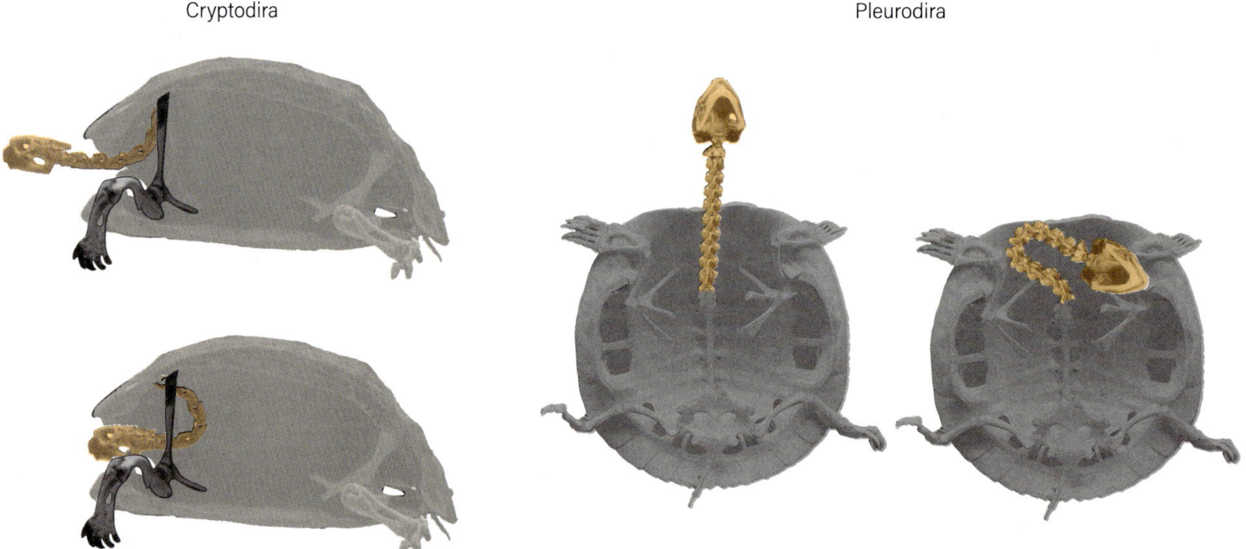

America. If you live in Canada or the United States, these are almost certainly the turtles you are familiar with.

There are approximately 280 species of hidden-necked turtles worldwide. For the purposes of this book, we can divide the hidden-necked turtles into six groups, omitting just a few families that are not found in North America. The diagram below illustrates all the major groups of North American turtles.

Canada and the United States are home to 64 turtle species, including representatives from all of the major groups of hidden-necked turtles shown below. All of these species are covered in the Species Accounts section of this book.

COMMON ANCESTOR CLADOGRAM

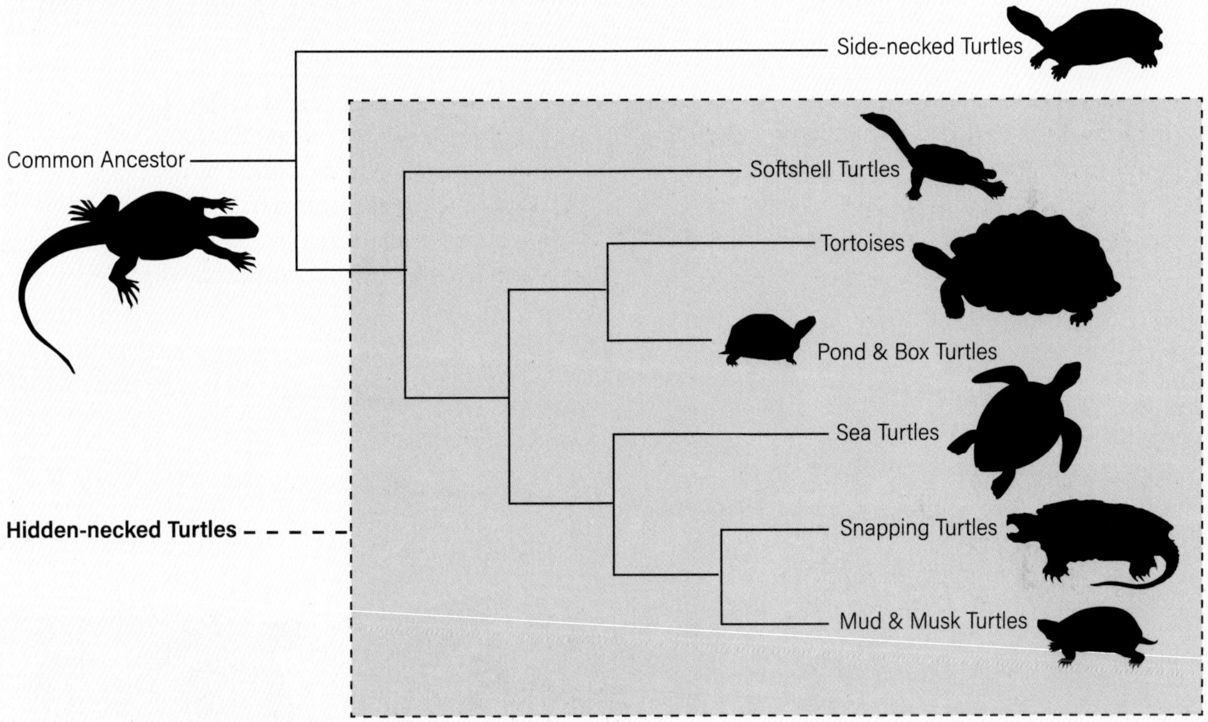

This cladogram shows the six groups of hidden-necked turtles that are native to Canada and the United States. Looking at the lines in the cladogram will tell you how these groups are related to each other, and to the side-necked turtles. If the lines between two groups are short, those groups are closely related to each other. The longer the lines, the more distantly related the groups are.

TURTLE ANATOMY

THE SHELL

Nothing distinguishes a turtle from other animals more than its shell. This most prominent turtle feature acts as a suit of armor that protects the turtle from the teeth and claws of its predators and, in some cases, even the clumsy feet of the larger animals that live around it.

All turtles have a shell, though the shells vary in size, shape, structure and texture. Some shells enclose the turtle completely in a protective box, while others barely cover the head and legs. Many are hard and impenetrable, but some are surprisingly soft and flexible. This seemingly simple covering is integral to a turtle's life, but many people don't fully understand how it works.

Contrary to some popular depictions, a turtle's shell is not something that the animal lives in. It is a part of a turtle's body and is integrated with the turtle's skeleton, organs, and skin. It is one of the most extreme specializations of any vertebrate animal and affects every aspect of a turtle's life.

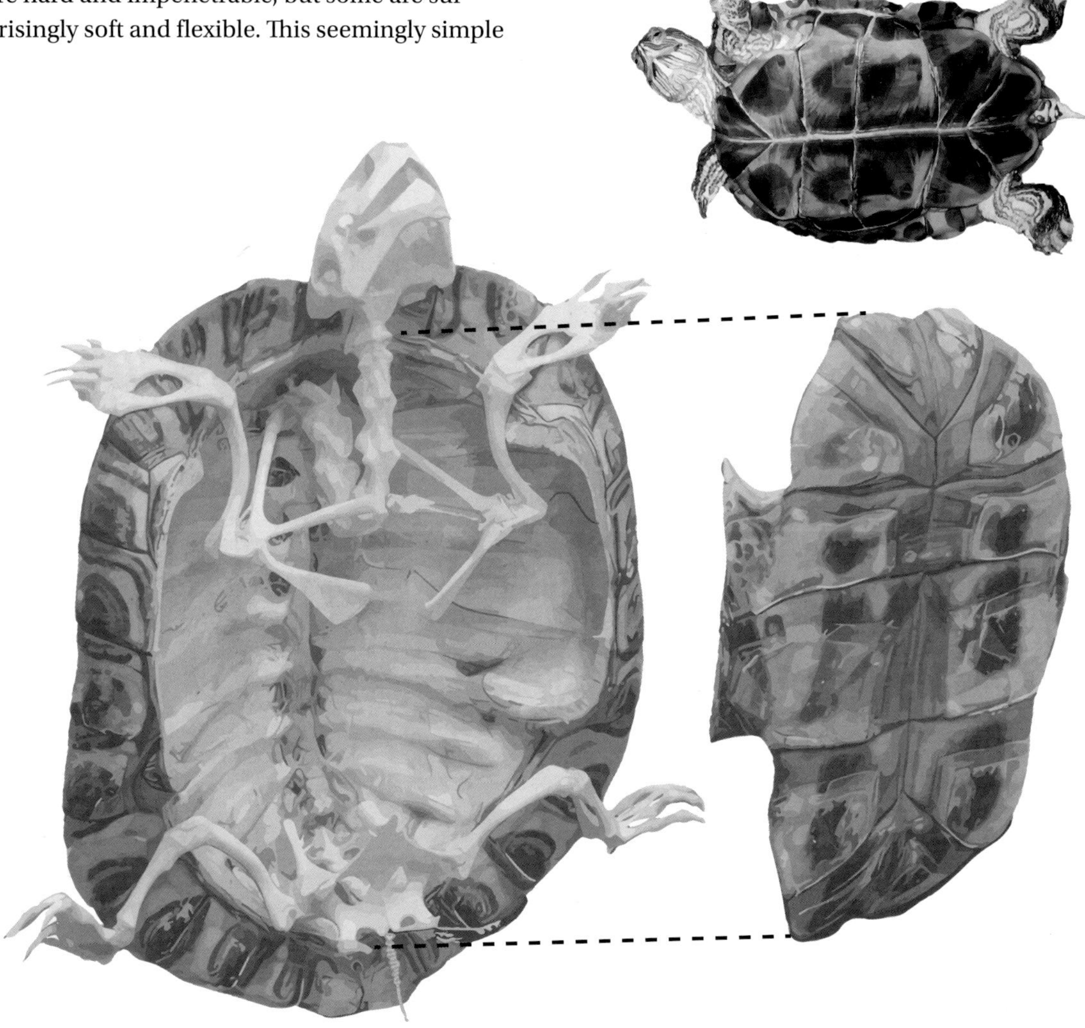

The shell is made of both bone and skin and it is attached to the other bones in the turtle's body. The upper part of the shell — called the carapace — is fused to the spine and ribs. The lower part — called the plastron — is attached to the bones that support the legs and includes the turtle's sternum. The carapace and plastron are joined at the sides, leaving gaps at the front and back to accommodate the turtle's head, legs and tail.

This heavy, rigid shell with its attachments to the skeleton help keep a turtle safe, but they also make the turtle's body stiff and inflexible. This is the reason that turtles are so famously slow, and the shell represents a major compromise between safety and speed. Most turtles have evolved to be slow and secure, but there are a few exceptions to the rule.

The bodies of reptiles are covered in scales, and a turtle's shell is no different. The skin that covers the shells of most turtles forms large, flat scales called scutes. These scutes are made of keratin, which is the same substance that makes up your fingernails and hair. Turtles shed their skin just like other reptiles, and you may sometimes see the worn scutes of a turtle's shell flaking off to reveal fresh, new scutes underneath.

The shape of a turtle's shell depends on the demands of the life it leads. Turtles who live mostly on land, like tortoises and box turtles, often have thick, high-domed shells. This heavy armor provides protection from land-based predators and the rounded shape helps prevent crushing if the turtle is stepped on by a larger land animal. These terrestrial turtles are often especially slow, but speed is a reasonable sacrifice if you're built like a tank.

Turtles that spend time both on land and in the water, like many of our common pond turtles, typically have shells which are flatter than their land-living cousins, but still rounded enough to offer some crush-protection. Shells with a flatter profile slip more easily through the water, making these turtles better swimmers. These lighter, slimmer shells also make the turtles a bit more nimble on their feet, allowing them to dash more quickly

The Gopher Tortoise (**ABOVE**) and Spiny Softshell (**BELOW**) represent two extremes in shell struture.

Photo © Kyle Horner

Out with the old, in with the new. A Northern Map Turtle sheds its old scutes.

The high-domed shell of a Texas Tortoise is cumbersome but crush-resistant.

The slimmer shell of the Pond Slider provides both protection and a good profile for swimming.

The streamlined shell of a Green Sea Turtle is a hydrodynamic masterpiece.

into the water if approached by a predator on the shore.

Turtles that live almost entirely in the water have the flattest shells of all. Sea Turtles, true swimming specialists, have flattened shells that taper at the ends. These sleek shells are perfectly shaped to be hydrodynamic, cutting through the water like an airplane wing slices through the air. These turtles may be more vulnerable on land, especially when they are small, but their swimming speed and efficiency are second-to-none.

There are many variations in turtle shells that help different turtles survive in different ways. The four turtles below illustrate some of the unusual ways that shells have become specialized.

SPECIALIZED SHELLS — BOX TURTLES

The box turtles live mostly on land and they share their high-domed carapace with the land-loving tortoises. To see the most unusual feature of a box turtle shell, though, you need to turn the turtle over. The box turtle's plastron is hinged both at the front and back, so the turtle can push the edges of the plastron up against the margins of the carapace. This creates a complete seal around the entire shell, and fully protects the turtle's head, legs, and tail inside an impenetrable box.

A male Common Box Turtle in box mode. Photo © Kyle Horner

The high-domed shells of box turtles, tortoises and some other species require one more special feature. The tall, unwieldy shape of these shells makes mating, where the male must climb on top of the female, rather difficult. To facilitate this important task, males of many species have a concave, or indented, plastron. The shells of the males and females fit together like puzzle pieces, allowing them to mate with relative ease.

SPECIALIZED SHELLS — SNAPPING TURTLES

The snapping turtles are best known for their readiness to bite when threatened, but to understand this seemingly ferocious behavior, you simply need to look at a snapper's shell. A snapping turtle's plastron is very small, and it doesn't protect the turtle's large head, legs, and tail. The small plastron gives great mobility to the turtles' legs, making them excellent walkers, but it also leaves snapping turtles vulnerable to predators like raccoons and foxes.

Since snapper shells are not a good defense, the turtles often resort to offense when encountered. This may make them intimidating, but if given space, snapping turtles will always seek safety over

A Spiny Softshell sporting its namesake feature.

conflict. The musk turtles share the snappers' small plastron and propensity to bite, though their tiny size makes them somewhat less threatening.

SPECIALIZED SHELLS — SOFTSHELL TURTLES

As its name suggests, a softshell's shell is a little different than that of most turtles. The shell is still bony in the center, but the surface has soft, leathery skin in place of the usual hard scutes. The bone does not extend all the way to the edges, so the outer margin of the shell is supple and flexible. These shells are also extremely flat in profile, and children seem almost instinctively to refer to these creatures as "pancake turtles."

These may seem like silly adaptations for a turtle, but they serve the softshell well. These turtles spend most of their time in water, and their flat, flexible shells combine with their strong legs and large, webbed feet to make them fast, powerful swimmers. These features also make them surprisingly fast runners over short distances on land, and with no hard shell for protection, the ability to make a mad dash back to the water is key to their survival.

A Snapping Turtle's tiny plastron, and the reason for all the snapping.

The bizarrely shelled Leatherback Sea Turtle.

SPECIALIZED SHELLS — THE LEATHERBACK SEA TURTLE

The Leatherback Sea Turtle surely has the most unusual shell of any turtle. Like the softshells, it has no covering of hard scutes. Unlike the softshells, it also lacks the solid, bony layer that gives all other turtle shells their structure. Instead, the Leatherback's shell contains a network of small, bony deposits called osteoderms, which are flexibly connected together.

The Leatherback's bizarre shell may provide less protection than that of other turtles, but Leatherbacks are so large that this often does not matter. Instead, this shell can flex and bend to deal with the expansion and contraction caused by ocean pressures, allowing this turtle to dive to depths of over a kilometer in search of the jellyfish it eats.

HEAD AND NECK

Turtle shells are hard and rigid, so many species compensate for this lack of mobility with a long, flexible neck. Their ability to extend and bend their necks in all directions allows them to lunge forward towards a tasty morsel of food, snorkel upwards to breathe at the water's surface, and even defend themselves by reaching out and biting an attacker.

At the end of that long neck, a turtle's head is compact and sturdy. The skull is stout and solid, built for durability and a powerful bite. Strong muscles at the back of the skull power the jaws, which are not equipped with teeth but with a

The Blanding's Turtle is particularly gifted in the neck department.

The ears of a Red-eared Slider are not red at all. The actual ear is a smooth, round membrane located just below the deceptive red stripe.

The nose of a Snapping Turtle, or a sophisticated carrion-detection device.

simple beak made of keratin, the same material as the outer layer of the turtle's shell. The beak can be sharp for slicing, flat for crushing, or even serrated like a saw for shearing meat or vegetation.

While no turtle has teeth, some have tooth-like bony projections inside their mouths that help them grab, hold, and swallow their food. The most extreme example of this is the Leatherback Sea Turtle, which eats mostly jellyfish. This slippery, squishy prey is hard to hold on to, so the Leatherback's entire mouth and throat is lined with sharp, backward-facing spines. Once a jellyfish is sucked in, the spines prevent it from slipping out and allow the turtle to slowly move it down towards the stomach.

THE SENSES

If you are looking for turtles in the water, the head of a swimming turtle may be the only part you see. Many aquatic species rest with their entire body submerged, and only the eyes, ears, and nose poking up above the surface of the water. In this way the turtle stays hidden, but can still scan its environment for food, predators and potential mates.

The eyes of aquatic turtles are often situated high on the head, to facilitate this sort of sneaky peeking at the world around them. This ability is less important for turtles who live on land, and their eyes are often correspondingly lower. Regardless of where they live, most turtles have excellent vision both above and under the water. They typically see color well and are excellent at spotting movement. Anyone who has tried to sneak up on a basking turtle can attest that the slightest motion is easily detected and triggers an immediate plunge to safety.

Located just behind the eyes are a turtle's ears, though they are sometimes quite difficult to see. The ears have no opening, they are simply exposed eardrums which look very similar to the rest of the turtle's skin. Turtles hear well and their unusual ears also work underwater, where they are particularly attuned to low-frequency sounds.

Most turtles have a great sense of smell, which they use on land and in the water to detect predators and search for food. Many turtles feed on carrion when the opportunity arises and there could hardly be a better sense to use when locating decomposing meat. Some turtles also use their noses to navigate through their environment and males of some species use scent to find females during the mating season.

Even the intimidating Snapping Turtle has no teeth.

The throat of a Leatherback Sea Turtle is a jellyfish's worst nightmare.

A Southern Painted Turtle cautiously surveys its surroundings.

LEGS AND FEET

As turtles live both on land and in the water, their legs and feet represent a compromise between the demands of walking and swimming. Some species need to do both equally, while others are more specialized and use one method of locomotion over the other. We can categorize turtle limbs into three main types, based on what type of movement they are built for.

Box turtles and tortoises have legs and feet which we could describe as elephantine. These legs are sturdy, stocky and upright, often with very short toes. Elephantine limbs are perfect for slow and steady walking, which is exactly how tortoises, box turtles and elephants get around. In many tortoises the front legs are also heavily armored with thick scales. The legs can be pulled in front of the face to shield it from predators.

Elephantine legs are great for walking, but very bad for swimming. Their upright position and small toes with minimal webbing give very little propulsion in the water. Many turtles with elephantine legs also have heavy, blocky shells. Box turtles

A Sonoran Desert Tortoise does its best elephant impression.

can swim clumsily for short distances, while most tortoises cannot swim at all.

Turtles which need to both swim and walk have legs that are splayed to the side, allowing a much more effective swimming stroke. They also have longer toes that are partially or fully webbed, which

The legs and feet of the Spiny Softshell aren't much for walking, but they are spectacular for swimming.

Photo © Kala King

The legs of the Green Sea Turtle are for swimming, and swimming only.

they use to push themselves through the water. There is variation among these species, some being better at walking and others better at swimming.

A Wood Turtle, for example, has legs that are only slightly more splayed and toes that are only slightly more webbed than those of a box turtle. Wood Turtles are capable swimmers, but often choose to simply plod along the bottoms of rivers and streams. On the other hand, the softshell turtles have widely splayed legs with long, fully webbed toes. They are poor walkers but exceptional swimmers, with all the speed and agility they need to catch the fish and other aquatic animals they eat.

The sea turtles, with little need to walk, have perhaps the most unusual legs of all. The legs and feet of a sea turtle are modified into flippers, which have no separated toes and are splayed completely to the side. These appendages are entirely specialized for swimming, with the oar-like front legs powering the turtle forward and the rudder-like back legs steering the animal through the water.

Flippers are not good for walking so sea turtles are extremely ungainly on land. Male sea turtles never leave the water, but females must come on shore to lay their eggs. They are unable to lift their bodies off the ground and must use their flippers to simply drag themselves onto their nesting beaches.

CLAWS

Regardless of the type of legs and feet a turtle has, nearly all turtles have claws. Even sea turtles, except for the Leatherback Sea Turtle, have one or two claws on their front flippers. The claws of some species, like the snapping turtles, can be particularly large and intimidating.

Turtles use their claws to help them walk on land or push along the bottoms of the lakes and rivers they live in, but for most turtles claws serve another vital function: they enable the turtle to dig. Turtles most likely evolved from burrowing ancestors, and that family history is especially apparent in the tortoises, many of whom dig burrows in which to live. Box turtles dig too, and while they cannot produce the perfect burrows of the tortoises, they often bury themselves in soil and leaf litter for protection.

Even turtles that don't burrow still rely on their digging abilities at a crucial stage in their lives.

A well-armed Snapping Turtle.

The long, dainty claws of the male Painted Turtle.

Almost all turtles lay their eggs in sand or soil, and they use their hind feet to excavate and then cover up these nest holes. Digging is vital to a turtle's life, even if some species only do it on very rare occasion.

Many turtles also use their claws when eating, especially to break up large food items. The Snapping Turtle, for example, loves to eat carrion, and will happily feast on dead animals much bigger than itself. Upon finding a large carcass, the turtle will bite with its powerful jaws and use its front claws in a prying motion to tear off chunks small enough to swallow.

In a few species, the claws play an unusual role in mating. The male Painted Turtle has very long, straight claws on his front feet. When he meets a female, he rapidly vibrates these claws in front of her face. If she is impressed, she does the same to him. Once this dance is complete she may allow him to mate.

THE TAIL

Turtles may have the least remarkable tails of all the reptiles, as most turtles have tails which are short and nondescript. The snapping turtles are the only North American exception, with long tails augmented by dinosaur-like spines.

Just under the base of the tail is the turtle's cloaca, which is the opening through which a turtle both poops and mates. The male's penis is stored inside the base of the tail and protrudes out from the cloaca when it is time to mate. Because they house the reproductive organs, the tails of male turtles are typically longer than those of females. The difference between the sexes is often small, though, and determining the sex of a turtle in the field by the length of its tail is usually not possible.

The tail of most turtles isn't much to write home about.

TURTLE BEHAVIOR

DIET

When it comes to eating, most turtles are anything but picky. The majority are opportunistic omnivores, meaning that they take advantage of whatever food is most readily available and eat both plants and animals. Some tend more towards carnivory (eating only meat), others lean towards herbivory (eating only plants), and a small number of species specialize and eat just a few specific foods.

Most of our pond, box, mud, musk, and softshell turtles are true generalists, and will happily eat some combination of insects, crustaceans, mollusks, soft-bodied invertebrates like worms, fish, algae, and plants. Many of these turtles only forage in the water, but those that eat on land — the box turtles and the Wood Turtle, for example — can also consume mushrooms, terrestrial invertebrates like millipedes, and plant parts such as flowers and berries that are rarely available in the water. The proportions of each food type eaten vary among species and depend on what is most available at any given time.

The snapping turtles, with their large size and powerful bite, can hunt larger prey. This may include fish, frogs, snakes, other turtles and even small mammals or birds. They have a notorious reputation for preying on ducklings at the water's surface, though this happens much less frequently than anecdotes often suggest. These seemingly fearsome turtles also eat plenty of invertebrates, and a surprising amount of vegetation.

Nearly all turtles readily scavenge carrion when it is available. Snapping Turtles, in particular, can often be seen gleefully feasting on dead animals of all sizes and in varying degrees of decomposition. A large carcass in a wetland is cause for great excitement and may attract multiple turtles of

A Common Box Turtle takes advantage of some fallen serviceberries.

The Gopher Tortoise lives for salad.

They may not look appetizing, but sponges seem to be the Hawksbill Sea Turtle's favorite meal.

various species. Scavenging turtles play a vital role in keeping wetlands clean.

The North American tortoises depart from the omnivorous lifestyle, as they have an almost exclusive taste for plants. Grasses, cacti, leafy plants, flowers, and fruit make up most of their diet, which is only rarely supplemented with insects or carrion. Several cooters, too, are primarily vegetarian as adults, though young cooters eat a variety of invertebrates as they are growing.

With their oceanic lifestyle, the sea turtles have a drastically different choice of food than terrestrial and freshwater species. Most feed on a wide variety of marine invertebrates such as crustaceans, mollusks, jellyfish, anemones, sea stars, squid, worms

and sea cucumbers. Fish and fish eggs are also eaten, as are various forms of algae (or seaweed) and even plants that grow along the seashore.

The Hawksbill Sea Turtle is somewhat more selective, as the majority of its adult diet consists of sea sponges. Sponges don't move and would appear to be easy prey, but their soft bodies are studded with glass-like spines and many are extremely toxic. The Hawksbill is seemingly unbothered by these defenses and, at times, sponges may constitute up to 95% of its diet. The Leatherback Sea Turtle, too, is a specialist, consuming almost exclusively jellyfish to sustain its massive bulk.

RESPIRATION

All turtles, even those that live their entire lives in the water, need to breathe air. A Leatherback Sea Turtle may dive to ocean depths of over a kilometer for durations of longer than an hour, but between dives it must poke its head out of the water to take oxygen into its lungs.

All turtles have lungs and their need to breathe air influences the way they behave. Many aquatic turtles float just under the surface of the water with only their eyes and noses protruding. This position keeps them as hidden as possible while allowing them to breathe easily, and prepares them to take a quick breath and dive at the first sign of danger. Observing turtles in the wild is sometimes a matter of picking protruding turtle heads out amongst the rocks and vegetation in a pond or wetland.

Unlike fish, no turtle has gills for extracting oxygen from the water. Some turtles, though, have evolved unusual strategies for getting small amounts of oxygen from their aquatic environments. Some species can absorb oxygen directly through certain areas of their skin, in a process called cutaneous respiration. At least one species, the Eastern Musk Turtle, uses a specialized tongue to draw oxygen from the water. The most bizarre method for underwater breathing, however, involves another part of the anatomy altogether.

Some aquatic turtles use a technique called cloacal respiration, in which the turtle sucks water

A Hawksbill Sea Turtle sneaks a breath between dives.

Unusually, the Eastern Musk Turtle absorbs oxygen through its tongue.

into its cloaca (the opening at the base of its tail). The water enters two special sacs inside the cloaca, which are lined with many finger-like projections. These tiny fingers absorb oxygen from the water and move it into the turtle's circulatory system. There's no way around it: these turtles breathe with their butts.

None of these methods allow a turtle to stay underwater indefinitely, but they extend the amount of time the turtle can remain submerged. During hibernation, when a turtle's body systems are drastically slowed down, underwater respiration may provide all the oxygen a turtle needs. When spring arrives, however, the turtle must again return to the surface to breathe fresh air.

This Painted Turtle could be butt breathing, and you wouldn't even know it.

A group of Pond Sliders gathered for a spot of sunbathing.

THERMOREGULATION

Like many of their reptilian cousins, turtles cannot produce their own body heat. For this reason, turtles and other animals like them are often referred to as cold-blooded. This unfortunate term suggests that turtles are always cold, but that is far from the truth. At times, turtles may have a body temperature higher than that of a mammal, but how they maintain and adjust that temperature is what makes them different.

Mammals, including humans, manage their body temperature from within. We convert energy from the food we eat into heat, just like the furnace in your house uses fuel to produce warm air. Our brain is like the thermostat that controls the furnace, and it makes constant adjustments to keep the temperature exactly right. Animals that use this strategy are endothermic, a term which literally means "inside heat."

Turtles are ectothermic. This term means "outside heat" and gives us a hint about how turtles do things differently. A turtle cannot adjust its body temperature from inside, so it must use its environment instead. If a turtle needs to warm up, for example, it may swim to warmer water or bask in the sun. To cool down it can retreat to the shade or take a cool plunge to the bottom of a lake or river. By moving around in its habitat, a turtle can maintain a relatively consistent body temperature.

Some turtles are so large that it takes a long time for them to warm up or cool down. The biggest turtles use this to their advantage. A sea turtle may bask at the surface of the water, warming its

A leisurely soak helps an Eastern Box Turtle cool off on a hot day.

body in the sun, then dive deep in search of prey. While diving in the frigid depths, the turtle's body begins to cool, but its great size allows it to retain the warmth from the sun for longer than a smaller turtle could. Even on hour-long dives, a sea turtle can stay warmer than the water around it, allowing it to hunt actively before it returns to the surface to bask once more.

A turtle's need to thermoregulate, or adjust its body temperature, can help us understand its behavior. For example, a hot, sunny afternoon may be a bad time to search for a Desert Tortoise, as it will likely escape the heat by resting deep in its burrow. A Painted Turtle, though, may take advantage of the warm sun to crawl out of the cool water and bask on a log. A cold, rainy day will likely be a poor occasion to find turtles of any sort, as they have no way to warm themselves and might remain inactive until the sun emerges.

OVERWINTERING

Because ectothermic animals, like turtles, do not produce their own body heat, they cannot remain active during the winter in most parts of Canada and the United States. Northern turtles, then, need a winter survival strategy. For nearly all of them, that strategy is hibernation.

As temperatures begin to drop in the autumn, turtles move to their hibernation sites. Most aquatic turtles will spend the winter in the mud at the bottom of a body of water. Here the temperature of their bodies will match that of the surrounding water, staying just a few degrees above freezing. Their heart rates may slow to less than one beat per minute, and most will get what little oxygen they need from cutaneous or cloacal respiration. Their body systems work so slowly that some can even survive without any oxygen for part of the winter.

A Common Box Turtle's hibernation strategy may seem a bit rudimentary, but it works.

The Leatherback Sea Turtle — a wanderer at heart.

You might imagine that, like a hibernating mammal, a turtle in this state would be fast asleep, but they are not completely inactive. Even in nearly freezing water, a turtle may move around to find the location with the best temperature or oxygen level. The sight of a turtle moving under the ice can be alarming indeed, but it is a natural part of their hibernation.

Some turtles hibernate on land. Many tortoises simply retreat into the security of their burrows to wait out the cool months. The North American tortoises all live in relatively warm environments, so they do not need to survive long periods of freezing temperatures. Some aquatic turtles in warm areas — like the Northwestern and Southwestern Pond Turtles, Chicken Turtle, and some mud turtles — will also hibernate on land, simply by burying themselves in soil and leaf litter.

Land-dwelling Common Box Turtles in the northern parts of their range face winter temperatures that can dip below freezing for much of the season. The box turtle cannot burrow like a tortoise, so it digs only a few inches into the soil. This shallow burial leaves it exposed to cold temperatures, but when the icy chills of winter arrive, the box turtle does something remarkable: it freezes.

For short periods, box turtles can allow over half of the water in their bodies to freeze solid. Then they simply wait out the freezing period and thaw when the temperature warms again. They produce large amounts of a simple sugar called glucose inside their bodies that acts as a sort of antifreeze, preventing ice from damaging the delicate tissues of their vital organs. This incredible ability is shared by some baby turtles who overwinter in their nests.

Hibernation is a vital part of life for most northern turtles since, at turtle speed, migration is not usually an option. The sea turtles are the exception, as they have the ability to swim great distances in the open ocean. This is especially true of the Leatherback Sea Turtle, which may wander as far north as Newfoundland in the summer, then migrate to South American waters in winter, making the longest migration journey of any reptile.

PREDATORS AND DEFENSE

Turtles, being mostly slow and ungainly, certainly don't have a reputation for having a fearsome or ferocious nature. On the contrary, they are truly defensive specialists. Most adult turtles have few predators, and once a turtle reaches adulthood, it has a good chance to live a long life.

Turtles are cautious animals by nature, something that anyone who has tried to sneak up on one can attest. Aquatic turtles are always ready to dive into the water at the slightest sign of danger and will often remain hidden in the muck and mire until long after the threat has passed. Even a tortoise will try to make a hasty retreat to its burrow, although haste is far from its forte.

Photo © Kyle Horner

This Wood Turtle is missing a leg, but it doesn't seem to cause her much trouble.

If pressed, many turtles will try to bite or scratch and their strong beaks and sharp claws may make a would-be predator rethink its dinner selection. The snapping turtles, with limited ability to retract into their shells, rely on this strategy more than most. Most species, though, depend heavily on their in-built suit of armor to keep them safe in dangerous situations.

Most turtle shells are sufficiently impenetrable to thwart likely predators. Particularly persistent hunters like crows, otters, raccoons, and some birds-of-prey may have success picking at the exposed parts, or even cracking the shells of smaller turtles. It is not uncommon to see turtles in the wild missing a leg or even two, although they seem to function surprisingly well without them. In the American southeast, alligators and crocodiles have strong jaws that make short work of all but the largest turtles.

Adult turtles are rarely eaten, but turtle eggs and hatchlings are food for a wide array of other creatures. Many predators, like raccoons, foxes, skunks, and crows, eagerly dig up turtle nests to gorge on the nutritious eggs inside. Even if they

Photo © Kyle Horner

The Eastern Musk Turtle may be small, but it's always ready to bite when it feels threatened.

The awesome American Alligator is one of few predators that can crack a large, adult cooter.

An industrious American Crow digs up a turtle egg for breakfast.

A Great Egret makes a quick meal of a young turtle.

survive incubation and emerge from the nest, the tiny hatchlings will be vulnerable to the same predators, as well as fish, large frogs, snakes, various birds, weasels, rodents, and even other turtles.

The odds are certainly stacked against these eggs and hatchlings, but turtles compensate for this by laying hundreds or even thousands of eggs over their lifetimes. While the vast majority will be eaten within the first year, only a few need to survive to replace the adults who produced them and keep the population stable.

COURTSHIP

The business of finding a mate is different for different turtle species. In some, like box turtles, tortoises, and snapping turtles, the males are larger than the females on average, and fight for access to territory and mates. This fierce combat may include chasing, ramming, pushing, biting, and even flipping. The contests can be quite long, only ending when one male retreats or is flipped over. Winning males may mate with females, whose smaller size means they often have little say in the matter.

In many other species the males are smaller than the females. This difference in size is especially dramatic in many pond turtles. In the Northern Map Turtle, for example, males weigh up to 0.4 kg (0.9 lbs) while females may reach 2.5 kg (5.5 lbs). Being so outmatched by their potential mates, males of these species focus less on fighting with each other and more on competing for the female's attention and approval.

Males of some species perform elaborate courtship displays to win the affection of a female. These demonstrations often occur face-to-face,

Snapping Turtles either fighting or mating. Who can tell?

Photo © Kevin Ricker

Long fingernails are the key to romance for the male Painted Turtle.

Mating in Common Box Turtles is cumbersome at best.

and involve some combination of head-bobbing, face-stroking, and gentle biting. Males of some pond turtles have specially elongated front claws which they vibrate rapidly in front of the female's face. Males may also release chemical signals called pheromones to attract females.

Whatever the method, once the courtship is over the male must mount the female. This is especially tricky for species with high-domed shells. Males of these species often have concave or indented plastrons which fit over the carapace of the female. Even with this adaptation, it is never a graceful affair. Turtles are persistent, though, and when everything is lined up correctly, the male inserts his penis into the female's cloaca and the deal is sealed.

While some turtles have specific mating seasons, in many species it can occur throughout the year. Female turtles can store the sperm from their mates until the time is right for egg-laying. They may mate with multiple males, storing sperm from each. Most species lay eggs in the spring or early summer, when temperatures are warm enough to incubate the developing young.

NESTING

Neither turtle parent cares for the young. The males typically leave immediately after mating. The eggs grow inside the body of the female and, when they are fully developed, she must find the right place to lay them.

Even turtles that live their lives in the water must lay their eggs on land. Choosing exactly the right nest site is vital, as the eggs must incubate at the right temperature and the right humidity. For most species, this means the female must lay them in a hole dug in soft, sandy soil. A few species skip the digging and simply lay the eggs in vegetation or a natural crevice.

The quest for the perfect nesting location can take these turtles on long and perilous journeys over land. Female Blanding's Turtles, for example, may walk well over two kilometers from their wetland homes to find the right nesting site. At turtle speeds, this journey may take over two weeks.

A Northern Map Turtle, having selected just the right spot, digs a nest with its hind feet.

These travels leave them vulnerable to predators and other dangers but choosing the right spot to lay their eggs is of the utmost importance.

Sea turtles are unrivalled in the scale of their journey to find a nesting site. A Green Sea Turtle, for example, may migrate over 2,000 km (1,250 mi) through the open ocean to return to the specific beach on which it hatched. There it will haul itself onto the sand to lay its eggs before returning to the ocean and beginning its migration back to its feeding grounds.

Once a female turtle has found a suitable nesting site, she excavates a nest with her hind feet and lays her eggs. The eggs may be round or oval, and in most species have a somewhat soft, leathery shell. The number of eggs depends on the species, with some tortoises laying only a few, and some sea turtles laying as many as 200. Once the laying is complete, the female turtle fills in the hole and departs.

SEX DETERMINATION

In almost all animals, the sex of the offspring depends on their genes. Humans, dogs, birds, frogs, and even insects receive chromosomes from both of their parents and the combination of those chromosomes determines whether the offspring will be male or female. This is called genetic sex determination.

Some turtles do this too, but others use a more unusual system. In many species, the sex of the offspring is determined by the temperature at which the eggs incubate, in a mechanism called temperature-dependent sex determination (TSD). In TSD, a turtle's genes have no bearing on whether that turtle will be male or female.

Complicating things further, there are two different types of TSD. In Painted Turtles, for example, lower incubation temperatures will produce males and higher temperatures will produce females. This is called Pattern I. In Eastern Musk Turtles,

A Green Sea Turtle deposits its eggs in a nest on the beach.

however, low temperatures produce females, intermediate temperatures produce males, and high temperatures produce females. This is called Pattern II. The exact temperatures at which the sexes switch varies between species and both patterns of TSD are relatively common in North American turtles.

HATCHING

The length of time that the eggs must stay in the nest varies widely, depending on both the species of turtle and the temperature of the nest. For most turtles incubation takes roughly between two and three months, although there are some both shorter and longer. The baby turtles typically hatch in the late summer or early fall.

A hatching Hawksbill Sea Turtle with an egg tooth on the tip of its nose.

A clutch of brand new Snapping Turtles emerges into the world.

Photo © Phyllis Holst

Escaping from an egg requires a special tool to cut the shell. A hatchling turtle is equipped with an egg tooth, a hardened point on the tip of its face that is used to slice through the eggshell and allow the turtle to make its way into the world. The egg tooth serves no other purpose, and the turtle loses it soon after hatching.

Once free from the egg, the tiny turtle must dig its way up through the filled-in nest to the surface. Once on the surface, hatchling turtles are extremely vulnerable to predators like birds and raccoons, so they must make their journey to the relative safety of the water as quickly as possible. In terrestrial turtles, like box turtles and tortoises, the young find shelter in vegetation, leaf litter, or natural cavities.

Some turtles, especially in cooler climates, may not dig out of the nest immediately. Rather than come to the surface as fall temperatures drop and hibernation looms, these turtles simply stay in the nest through their first winter. They do not eat, surviving only from the nourishment they received in the egg. Overwintering in the nest may expose these hatchlings to freezing temperatures, but some — including the Painted Turtle — are able to allow their bodies to partially freeze. When weather warms in spring, the young turtles finally leave the nest and head to the water for the first time.

A female Snapping Turtle laying one of thousands.

HATCHLING SURVIVAL AND LONGEVITY

The world is a challenging place for a baby turtle and most hatchlings do not survive to adulthood. Many predators, both on land and in the water, are waiting to gobble up these bite-sized morsels from the day they leave their nest. Turtles grow slowly, so for the first several years of their lives, their small size makes them vulnerable to the bigger animals around them.

Leaving your small, defenseless offspring to fend for themselves may not seem like a great strategy for survival, but for turtles, it's all about playing the long game. If a young turtle survives to adulthood, it has a very good chance to live for a very long time. Most adult turtles have few natural predators and turtle lifespans range from a few decades to over a century, depending on the species.

This longevity means that turtles may lay eggs every year for decades. Research suggests, for example, that a Snapping Turtle can live for more than 100 years. Female snappers begin to lay eggs at around 15 years of age and are able to continue to lay eggs every year for the rest of their lives. If a female Snapping Turtle lays 40 eggs per year for 85 years, she will have laid 3,400 eggs in her lifetime.

Many of those 3,400 eggs will be eaten by nest-raiding animals before they hatch. Of the turtles that do hatch, many will be eaten by predators before they are old enough to lay eggs themselves. None of this matters, though, because only a small number of these turtles need to survive to replace the long-lived adults, keeping the population healthy and stable.

OPPOSITE The dash to the water is a perilous journey for a baby sea turtle.

Turtles now live in a rapidly changing world.

TURTLE CONSERVATION

Although they are widely beloved, turtles are one of the most threatened groups of animals on Earth. Over half of the world's species are considered at-risk and many are critically endangered. With affection for turtles being so universal, it might be hard to imagine why they are in so much trouble.

The reason for this seems to lie in the way that the biology of turtles interacts with humans. Turtles have survived on this planet, relatively unchanged, for hundreds of millions of years. Their hard shells have protected them from predators and allowed them to live long, slow lives. Their longevity gives them time to lay hundreds or thousands of eggs, ensuring that a few of their offspring will survive to adulthood. They have adapted, over time, to live in whatever habitats are available to them, and those habitats have evolved and changed as slowly as the turtles themselves.

After millions of years of relative consistency, modern humans entered the picture and completely transformed the world that turtles lived in. In just a few thousand years we covered large parts of the planet with buildings, roads and farmland. We altered habitats, changed the composition of the water and air, moved species from one place to another, and hunted and ate many of the creatures that lived alongside us.

Some species have been able to adapt to meet the challenges. Some of the most adaptable — like the rat, pigeon, or raccoon — have even taken advantage of the novel opportunities that our presence provides. But turtles, who do almost everything slowly, may be uniquely poorly equipped to deal with the threats that they now face.

Turtles desperately need our help. Humans can certainly be a destructive force in nature, but we can do good, too. Some of the problems that turtles face today are complex and difficult to solve, but there are also ways to help turtles that are remarkably easy. Armed with a little knowledge, even young children can contribute to turtle conservation. We all have the power to make the world a better place for our shell-wearing friends.

The following six points are a summary of some of the biggest issues facing turtles today, along with some suggestions for actions that we can all take for the benefit of turtles.

1. HABITAT LOSS AND ALTERATION

Perhaps the biggest threat to most of our turtles is the loss or modification of the places where they live. This is especially true for the many species that live in freshwater wetlands. Swamps and marshes make excellent farmland and many such habitats have been drained to make way for agriculture. Often located beside larger bodies of water, marshes may also be filled in to build desirable housing or as a location for industry.

In southern Canada and the continental United States, over half of all wetlands have been lost since the first European settlers arrived. For the wetlands that remain, pollution and contamination may reduce their value as turtle habitat. Lower water quality means less vegetation, which robs turtles of both shelter and the foundation of their food webs. Toxins in the water may directly affect the turtles themselves. Healthy, productive wetlands are crucial for turtle survival.

The situation for turtle habitat in North America may seem dire, but there is reason for hope. Wetland destruction peaked just after the middle of the 20th century when new laws to protect wetlands and the species that inhabit them started to slow the rate of loss. People are starting to see the

The fertile soil and natural irrigation of wetlands makes them valuable areas for farming.

Healthy wetlands are vital to the survival of many turtle species.

Many hands make light work, and cleaning up nature is a way we can all help turtles.

value of wetlands for both wildlife and humans. A variety of organizations have arisen to advocate for wetland conservation.

For the individual, helping to conserve turtle habitats can be a daunting prospect. For those who are financially able, supporting charitable organizations that work to conserve wetlands may be a great place to start. Property owners can protect and restore wetlands on their own land, and even small tracts of habitat are extremely valuable. Voting for politicians and leaders who care about conservation costs nothing and is a fundamental way to make your voice heard.

Participating in volunteer projects like wetland cleanups, plantings, or restorations is a great way to get involved. Check with your local nature club or conservation organization to find these initiatives. For families and children, learning the value of proper waste disposal and cleaning up litter in your local marsh can be an educational and rewarding experience, and the time spent outdoors can help drive home the importance of these spaces for the animals that live there.

2. ROADS

For turtles, crossing a road is extremely dangerous. Not only are they unable to cross quickly, but millions of years of evolution have not prepared them

These new roadside decorations are cropping up everywhere.

Turtles and roads are a dangerous combination.

for the sudden development of motorized vehicles. Turtles simply do not understand roads, and they are unlikely to figure them out anytime soon.

This is a substantial problem, because roads have become an inevitable part of turtles' lives across North America. Our road network is so expansive that on much of the continent, it is impossible to walk a kilometer in any direction without crossing a road. They surround our wetlands and forests, slice through our deserts and trace the shape of the seashore.

For much of a turtle's life, it may not leave the small wetland or forest tract in which it lives. For one particular task, though, turtles make substantial journeys. That task is nesting. Female turtles may walk several kilometers over land to find the perfect place to lay their eggs and these turtles must cross roads.

When a female turtle is killed by a vehicle, the impact is far more substantial than the loss of just one turtle. Turtles have evolved to live long lives and lay eggs for many years. The survival of their young is very poor, so laying many eggs is an insurance policy to ensure that a small number of young turtles reach adulthood. If a female turtle is lost in her early years, she will never lay the hundreds or thousands of eggs needed for the survival of the species.

Road mortality is recognized as a severe threat to turtle populations and some solutions have begun to arise. Some turtle nesting hotspots throughout North America have been equipped with ecopassages: systems of fences and tunnels which allow turtles to cross under roads, unharmed. Turtle crossing signs have become a regular roadside feature near wetlands and campaigns to raise awareness of this issue are now common.

Individual action has an enormous capacity to help turtles with this threat. Simply keeping a careful watch for turtles while driving can help prevent the collisions that kill turtles. These slow-moving animals are easy to avoid, so if all drivers were on the lookout, many accidents could be averted.

Helping turtles cross the road can be extremely beneficial. Most turtles can be simply picked up and carried across the road in the direction in which they are moving. Once they are out of harm's way, the turtles will continue on their journeys, unaffected by their brief encounter with a helpful human. It is important to remember your own safety and to ensure you do not put yourself at risk while helping a turtle.

In many places, specialized wildlife rehabilitation centers exist to help turtles who have been the victims of vehicle collisions. Becoming familiar with the facilities in your area can prepare you for helping turtles that have been hit by cars. Even quite severe injuries can often be fixed

Photo © Kyle Horner

A Blanding's Turtle gets a helping hand to the other side.

by professional rehabilitators, so getting injured turtles to qualified help quickly is vital. Many such centers have websites with excellent information on the best ways to contain and transport injured turtles.

3. SUBSIDIZED PREDATORS

Turtle eggs and hatchlings are extremely vulnerable to predators and very few survive to adulthood. In a healthy ecosystem there is a natural balance between the turtles and the predators who eat them, and enough turtle hatchlings grow into adults to keep turtle populations stable.

Unfortunately, many turtle populations are declining and the populations of their predators — especially raccoons — are increasing. Doing your part to discourage raccoons can help keep inflated predator populations in check. Storing garbage in raccoon-proof containers and preventing access to outbuildings encourages raccoons to seek natural food and shelter rather than looking for an easy meal or hotel stay.

In many areas, initiatives are underway to reduce the impact of nest predators on turtles. In these projects, turtle nesting sites are monitored and cages are placed over the nests shortly after the eggs are laid. These cages prevent predators from excavating the nests, which are then closely watched so that the cages can be lifted when the turtle hatchlings emerge. These projects often need volunteers to locate and monitor nests, so finding one in your area can be a great way to get involved in turtle conservation.

4. COLLECTION AND HARVESTING

For some turtle species and in some parts of the world, the collection of turtles for food, pets, or even traditional medicine is a significant danger to wild populations. While eating turtles is not widely popular in North America now, it was historically common and still occurs today on a smaller scale. In many places the harvesting of turtles is illegal, although some states do allow hunting in small quantities with an appropriate license.

Larger-scale poaching of turtles does sometimes occur in North America, usually with the intent to ship turtles overseas where they are in demand as food. A turtle's long lifespan and slow reproduction make poaching extremely damaging to populations, as it may take decades to replace even a few turtles lost to collection.

While harvesting for food is relatively uncommon in North America, wild turtles are sometimes

Raccoons are almost too good at survival, and their relationship with turtles is out of balance as a result.

The Chicken Turtle was named for the taste of its meat, and was once extensively harvested for food.

taken as pets, either singly by would-be turtle owners or in numbers by those who wish to sell them for profit. The collection and sale of native turtles as pets is illegal in most places, but these laws are often difficult to enforce and some of our most at-risk turtle species are greatly endangered by demand from the pet trade.

For anyone considering a pet turtle, it is extremely important to ensure that the turtle you choose is responsibly and legally sourced. Turtles who are born and raised in captivity live healthier, happier lives as pets and you can rest assured when you acquire one that you are not impacting wild populations. In general, turtles are difficult animals to care for properly and a great deal of research should precede any decision to bring one into your home.

The adorable Bog Turtle is threatened by collection for the pet trade.

Once grown, these baby Red-eared Sliders will be more turtle than most pet-owners bargained for.

5. INVASIVE TURTLES

It may be counterintuitive, but sometimes the biggest threat to a turtle is another turtle. For decades, the Pond (or Red-eared) Slider has been sold in pet stores all over the world. These "dollar turtles" are undeniably cute as babies in a pet shop display, but they grow into large, active, messy turtles that are difficult to house and care for. Countless people have taken home a tiny turtle, only to find it too much to handle just a few years later.

Unfortunately, many well-meaning people have released these pet turtles into their local wetlands. While native to the southeastern United States, Pond Sliders are tough and adaptable, and have survived and thrived in many countries worldwide. These durable turtles can outcompete native

Ocean plastic is one of many threats to the world's sea turtles.

turtles for food and even push them out of their basking, nesting, and overwintering sites. They can also introduce diseases and parasites acquired during their time in captivity that are harmful to wild populations.

A pet of any kind is a lifetime responsibility. For turtles, that can mean decades of care. It is important to consider every aspect of pet ownership before deciding to purchase a turtle as a pet, and absolutely vital never to release a captive turtle into the wild.

6. SEA TURTLE CONSERVATION

The lives of sea turtles are drastically different from those of all other North American turtles, so logically the threats that they face are different too.

Almost all of the world's sea turtles are threatened, and their unique biology makes them especially vulnerable to certain dangers. Commercial fishing bycatch and marine plastics endanger turtles in the open ocean, while light pollution, beach modification, and poaching of both adults and eggs threaten them on their nesting beaches.

For most North Americans, sea turtle conservation seems remote from daily life and it can be difficult to see how our actions can affect these unique animals. There are organizations dedicated to sea turtle conservation all over the world and excellent online resources available for those who wish to learn more about how to help these incredible creatures.

SPECIES ACCOUNTS

The Species Accounts that follow will introduce you to the 64 species of turtles that live in Canada and the United States. The following is an explanation of the information you will find in each account.

SPECIES NAMES For each species I have tried to use the most commonly accepted common and scientific names, but turtle taxonomy is an evolving science and many changes have occurred in recent years with the advent of new techniques for examining turtle genetics. Other texts and resources may use different names or even different arrangements of species and subspecies. I have made every effort to indicate where such confusion may occur.

IDENTIFICATION This section provides characteristics that help to identify the species in the field. I have emphasized characteristics which can be observed at a distance and mostly excluded those which require up-close, in-hand examination. The latter are rarely useful in a field setting, and unnecessary capture of turtles for identification should be avoided. Some species are very difficult to identify in the field, but such is the challenge of turtles.

I have provided a maximum known carapace length for all species. Many texts use a size range or average size, but since all turtles begin their lives very small, it is best to assume that you could find a given species at any size between tiny and the recorded maximum. I have used carapace length instead of overall length, as the latter depends on the position of the turtle's head, legs, and tail. A frightened turtle in its shell has a much shorter overall length than one swimming through the water.

ALSO KNOWN AS Turtle names periodically change and many turtles have been historically known by more than one common name. I have provided here any additional names by which a species has been commonly known and that you may still find in other texts or resources.

SIMILAR SPECIES Here I have tried to list the turtles that you are most likely to confuse for a given species. Remember to check species' ranges when you are identifying turtles, as it can be an important factor in determining what species are possible wherever you are located.

RANGE Range maps are provided for all species, alongside a short, written description of their range. These maps are broad in scale and so should be considered somewhat approximate. Remember that each species has particular habitat preferences, so within the illustrated range the species will only occur where suitable habitat is present.

HABITAT I have given a brief description of the preferred habitat(s) for each species, along with notes on relevant behaviors. Habitat should be considered when identifying turtles in the field, but of course it is possible for a turtle to appear in an atypical habitat, so it is best not to exclude species on this basis alone.

DIET Most turtles are omnivorous and opportunistic, consuming a wide variety of different foods. I have attempted to indicate the food types that each species is known to regularly eat, but it should not be considered an exhaustive list, as a hungry turtle rarely refuses any potential meal.

REPRODUCTION Some details on the reproduction of each species are given here, specifically the maximum clutch size (number of eggs) the turtle is known to lay, the preferred nesting location and the maximum number of clutches per year that the species can produce. There are still gaps in our knowledge of the reproductive biology of some species and I have indicated where this is the case. For species which produce more than one clutch per year, the number of clutches is typically greater in the southern parts of their ranges where the warm seasons are longest.

SUBSPECIES Many turtle species can be broken up into smaller, distinct groups called subspecies. Often the subspecies live in different geographical locations or have physical differences that set them apart from each other. For each species, I have listed all the subspecies that occur in Canada and the United States and briefly described how they are differentiated.

CONSERVATION Nearly all North America's turtles are threatened to some degree and many are considered formally "at-risk." States, provinces, countries and international authorities all assess and designate threatened species in different ways, so it is possible for a given species to have different risk levels according to who is doing the risk assessment.

For each species I have provided its risk level as designated by the Canadian government (if the species occurs in Canada), the American government and the International Union for Conservation of Nature (IUCN). Each of these authorities uses a different set of designations, which are listed below. I have not included state and provincial designations, but these can be easily found on the web. For each species I have also indicated the most significant risk factors that threaten them in the wild.

CANADA

In Canada, at-risk species are protected by the Species At Risk Act (SARA). Many North American turtles have only small ranges in Canada, where the short summers and long winters create unique challenges to their survival. SARA designations only pertain to the Canadian population of a given species and it is possible for a species to be endangered in Canada even though it is common in the United States. There are four levels of risk that pertain to Canadian turtles:

SPECIAL CONCERN – the species may become threatened or endangered based on its biology and threats.

THREATENED – the species is likely to become endangered if action is not taken.

ENDANGERED – the species is facing imminent extinction or extirpation.

EXTIRPATED – the species no longer exists in the wild in Canada.

UNITED STATES

In the United States, species that are in trouble may become listed under the Endangered Species Act (ESA). The ESA specifically pertains to the population of a species in the United States. There are only two categories:

THREATENED – the species is likely to become an endangered species within the foreseeable future.

ENDANGERED – the species is in danger of extinction throughout all or a significant portion of its range.

GLOBAL

The International Union for Conservation of Nature evaluates the conservation status of species from all over the world. It publishes and maintains the IUCN Red List of Threatened Species, which lists species according to their global risk status beyond the borders of any one country. There are five categories of risk which pertain to the turtles in this book:

LEAST CONCERN – the species does not qualify for Critically Endangered, Endangered, Vulnerable, or Near Threatened.

NEAR THREATENED – the species does not qualify for Critically Endangered, Endangered, or Vulnerable now, but is close to qualifying or likely to qualify in the near future.

VULNERABLE – the species is facing a high risk of extinction in the wild.

ENDANGERED – the species is facing a very high risk of extinction in the wild.

CRITICALLY ENDANGERED – the species is facing an extremely high risk of extinction in the wild.

THE SNAPPING TURTLES

This family contains three large and infamous turtles that live in Canada and the United States. A fearsome reputation accompanies the words "snapping turtle," but as you get familiar with these misunderstood giants, you'll quickly see there is little cause for concern.

All of our snapping turtles grow to a considerable size. The two alligator snapping turtles are among the largest freshwater turtles in the world. Their hefty proportions are augmented by a blocky head, powerful legs and a long, spiky tail. Their overall appearance is truly prehistoric and these turtles have been relatively unchanged since dinosaurs walked alongside them.

The defining feature of the snapping turtles is perhaps not their powerful jaws, but their tiny plastrons. Their reduced belly armor gives their legs excellent mobility for walking and even occasionally climbing, but it also makes them wary of any creature bigger than themselves and quick to bite in defense.

Fortunately, snapping turtles are mostly inclined to bite on land where, being turtles, they are easily avoided. In the water, they rarely feel the need to lash out in defense. As enthusiastic scavengers and hunters of the easiest-to-catch animals around them, they play a vital role in keeping their native wetlands free from disease and decay. Snapping turtles do a dirty but important job and deserve our appreciation.

OPPOSITE Snapping Turtles are excellent swimmers, but you'd hardly know it. Walking along the bottom seems to be preferred whenever possible.

BELOW A Snapping Turtle's dirty, little secret exposed.

BELOW Walking while shelled is no easy task, but Snapping Turtles have the legs for it.

ABOVE The beaks of the snapping turtles look ferocious, but they are food-grabbers first.

THE SNAPPER'S SNAPPER

Like any mouth, this beak is built for getting food from the outside to the inside, and it does that with great efficacy.

Snapping turtles are hunters of a variety of small animals. They typically employ a sit-and-wait strategy, resting patiently on the bottom waiting for the perfect moment to lunge forward and clamp down on prey with their vice-like jaws. Once caught, an unsuspecting fish or frog has little chance of escape.

It may seem likely that hunting is the primary business of these formidable turtles, but their commitment to energy efficiency makes scavenging an appealing option. No dead animal is safe from a hungry snapping turtle and they will as happily feed on a mouse as a moose. To deal with large carcasses, a snapper simply grabs a mouthful of flesh in its beak and uses its sharp claws to rake and pry until a chunk comes free.

Snapping Turtles are shy creatures that only bite as a last resort. These prehistoric turtles absolutely love to eat carrion, making them important scavengers in their aquatic habitats.

IDENTIFICATION Carapace has a relatively flat profile with a serrated rear margin and three ridges on top that smooth out as the turtle gets larger. Color ranges from almost black to light brown or greenish. Plastron is small and yellowish or brownish in color. Head is large and blocky, legs are thick and powerful, tail is long and spiky. Carapace length up to 50 cm (20 in), weight may exceed 15 kg (33 lbs).

ALSO KNOWN AS Common Snapping Turtle

SIMILAR SPECIES Compare to the alligator snapping turtles. Musk turtles share their small plastron and readiness to bite in defense.

RANGE Widespread across eastern and central North America, from Nova Scotia west to the edge of Saskatchewan, south to New Mexico, the Gulf Coast and the southern tip of Florida. May occur outside of its native range where it has been introduced from captivity.

HABITAT Almost any body of freshwater including lakes, rivers, ponds, streams and marshes. Rarely basks out of the water. Often floats beneath the surface or sits on the bottom with just its head exposed. Occasionally wanders over land.

DIET Any small animal it can catch including fish, frogs, reptiles, birds, mammals and invertebrates. Frequently scavenges on carrion, readily eats plant material.

REPRODUCTION Lays up to 80 eggs in a nest excavated in soft, sandy soil, sometimes quite far from the water. Gravel roadsides are a frequent and

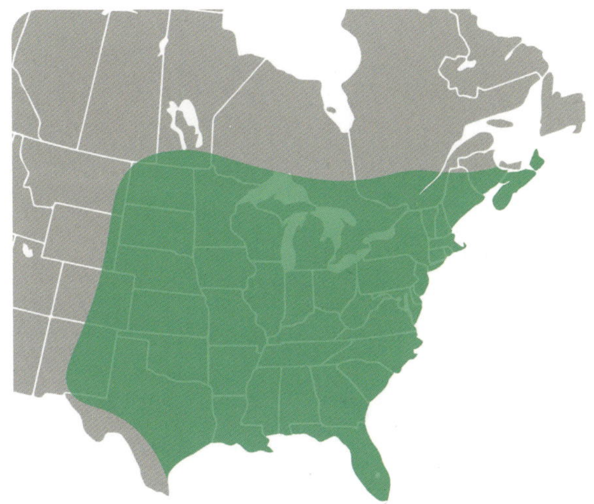

unfortunate choice for a nest site. Lays no more than one clutch per year.

CONSERVATION CANADA – Special Concern, USA – Not Listed, GLOBAL – Least Concern. Fairly common across most of its range but declining in some areas. Road mortality may be its biggest threat, but habitat alteration, harvesting and nest predation by subsidized predators also play a role.

SNAPPERS IN TROUBLE

Snapping Turtles have relatively small shells that don't fully protect their chunky heads, legs, and tails. This vulnerability makes them nervous on land, so they use their long necks and powerful jaws as their primary line of defense.

Crossing roads is big problem for these turtles and many are killed by cars as they walk in search of their perfect nest site. A helping hand across the road can be a real lifesaver, but how do you move a testy turtle that doesn't know it's being helped?

The best way to approach a Snapping Turtle is from the back. Lifting the turtle by its tail can injure it, so instead, grab the turtle firmly by the back of its carapace, just above its hind legs. Then gently lift, push, or drag the turtle to safety. Small snappers can simply be scooped up from the back. A Snapping Turtle's neck can reach about halfway down its shell, so be sure to keep your hands and face out of harm's way.

Always remember to move the turtle in the direction it was already headed, as it is on an important journey. By helping these misunderstood creatures navigate roads, you can help ensure that they are a feature of our swamps and marshes well into the future.

Photo © Kyle Horner

ALLIGATOR SNAPPING TURTLE
Macrochelys temminckii

The enormous Alligator Snapping Turtle is truly dinosauresque, and it's hard to believe a creature like this still exists in the modern world. These turtles rarely leave the water and so are rarely seen, spending most of their time lurking in the muddy depths of a river or wetland.

IDENTIFICATION Carapace is heavy with a relatively flat profile and has three prominent, spiked ridges along it. Rear margin is serrated. Color may be black, gray, brown, or greenish. Plastron is small and ranges from yellowish to brown or black. Head is massive and blocky, with a strongly hooked beak. Tail is long. Adults can be enormous, with carapace length up to 80 cm (32 in) and weight sometimes exceeding 80 kg (176 lbs).

ALSO KNOWN AS Loggerhead Snapping Turtle

SIMILAR SPECIES Nearly identical to Suwannee Alligator Snapping Turtle. Also compare to Snapping Turtle.

RANGE The Gulf Coast of the United States and its associated river systems, from the Florida panhandle to eastern Texas and as far north as northern Arkansas or southern Missouri.

HABITAT Deep river systems, lakes and wetlands, especially those with overhanging trees, undercut banks, dead wood and other cover. Mostly rests on the bottom, only occasionally moving to breathe or look for food. Rarely leaves the water.

DIET Mostly fish, amphibians, other reptiles and invertebrates. Occasionally birds, small mammals and plant material. Frequently scavenges on carrion.

REPRODUCTION   Lays up to 50 eggs in a nest excavated in loose soil near the riverbank. No more than one clutch per year.

CONSERVATION   USA – Proposed Threatened, GLOBAL – Vulnerable. A listing of Threatened under the Endangered Species Act of the United States has been proposed for this species. Historically hunted for meat. Now protected in most states, but populations have not fully recovered. Habitat alteration may now be the most significant threat.

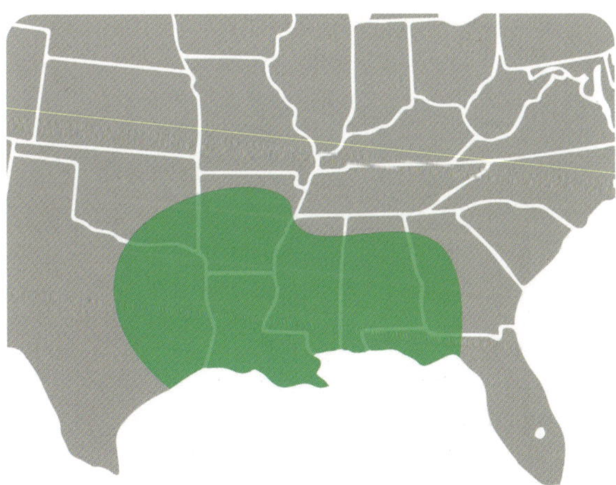

TONGUE TWISTER

It may look big and ferocious, but the Alligator Snapping Turtle uses a remarkably dainty tool for hunting. On the tip of its tongue, resting on the bottom of its cavernous mouth, is a little, pink appendage. It looks as if a worm crawled in there when the turtle wasn't looking, and looking like a worm is exactly what this little piece of flesh was designed to do.

When hunting for fish, the Alligator Snapping Turtle rests perfectly still on the bottom of a river or lake and opens its mouth wide. The dark colors and markings of the turtle allow it to blend perfectly into its murky surroundings but the bright pink tongue stands out amidst the muck and mire. When the turtle is ready, the tongue completes the illusion with one, final trick: it wiggles.

Fish are attracted to this bright, wiggling worm and come closer to investigate. If a fish ventures into range, the turtle snaps its hooked beak closed with incredible speed and power. There is no escape for the hapless fish. The turtle has snagged a meal using the absolute minimum amount of effort. A perfect demonstration of the old adage "work smart, not hard."

SUWANNEE ALLIGATOR SNAPPING TURTLE
Macrochelys suwanniensis

Until recently, the Suwannee Alligator Snapping Turtle was thought to be just a population of Alligator Snapping Turtles. Genetic research in the 2010s showed that it is a species in its own right. Because it is newly defined, limited information exists on this turtle that is independent from its sister species.

Photo © Grover Brown

IDENTIFICATION Very similar to Alligator Snapping Turtle in all respects, with only slight differences in the shape of the head and shell. Most easily identified by range, as the Suwannee Alligator Snapping Turtle only occurs in the Suwannee River basin, where its cousin, the Alligator Snapping Turtle, does not.

SIMILAR SPECIES Nearly identical to Alligator Snapping Turtle. Also compare to Snapping Turtle.

RANGE Occurs only in the Suwannee River basin in northern Florida and southern Georgia.

HABITAT Rivers and streams, and occasionally other wetlands connected to the river system.

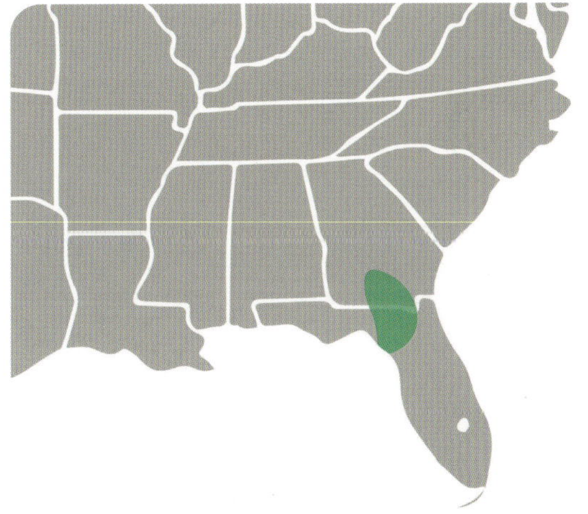

Photo © Grover Brown

Prefers areas with overhanging trees, dead wood, and other cover. Rarely leaves the water.

DIET Like the Alligator Snapping Turtle, hunts mostly smaller aquatic animals and forages for carrion underwater.

REPRODUCTION Little information exists, but likely similar to Alligator Snapping Turtle.

CONSERVATION USA – Proposed Threatened, GLOBAL – Not Assessed. A listing of Threatened under the Endangered Species Act of the United States has been proposed for this species. Historically hunted along with Alligator Snapping Turtle but now protected. Habitat alteration continues to pose a threat.

Family Kinosternidae
THE AMERICAN MUSK AND MUD TURTLES

- Eastern Musk Turtle
- Razor-backed Musk Turtle
- Loggerhead Musk Turtle
- Intermediate Musk Turtle
- Striped-necked Musk Turtle
- Flattened Musk Turtle
- Eastern Mud Turtle
- Florida Mud Turtle
- Striped Mud Turtle
- Rough-footed Mud Turtle
- Yellow Mud Turtle
- Arizona Mud Turtle
- Sonoran Mud Turtle

The musk and mud turtles comprise 13 of the smallest turtles in Canada and the United States, but what they lack in size, they make up for in attitude. These little turtles readily bite to defend themselves from animals (including humans) hundreds of times their size and can also employ a smelly musk to ward off attackers.

Most turtles in this group have relatively long, narrow, high-domed carapaces. Some species have ridges or keels which may help with identification. The musk turtles have very small plastrons, like the snapping turtles, which provide little protection. The mud turtles, on the other hand, have large plastrons that are hinged at both the front and back. These movable plates can be closed to some degree, although they don't typically create the tight seal of the similarly equipped box turtles.

Musk and mud turtles spend most of their time in the water, either walking along the bottom or floating at the surface where they are often concealed by vegetation. They hunt a wide variety of invertebrates, but are opportunistic and will eat other small animals, carrion, or plants if they are available. They rarely stray far from the water, even for nesting, which typically takes place quite close to the shoreline.

THE BEST DEFENSE IS OFFENSIVE

Musk and mud turtles are among North America's smallest turtle species. Their vulnerability makes them very defensive, and they are always ready to chomp an attacker if the occasion calls for it. Unfortunately, their ferocious attitude is somewhat undermined by their tiny size and biting alone may not ward off a bigger animal.

As an additional line of defense, a frightened musk or mud turtle can release a foul-smelling fluid from glands located on its sides, between its front and hind legs. The odor may invade the nose and mouth of a would-be predator and make the offender think twice about its next meal. Even a moment's hesitation may be enough for the turtle to scuttle away into the water, allowing this smelly creature to survive another day.

AN AIR OF MYSTERY

Some members of this group are among our most poorly studied turtles, and there are large gaps in our knowledge of their life history. Their small size, reclusive habits, and some taxonomic confusion has allowed them to fly under the radar. Recent research has just started to round out our understanding of this family. Genetic analyses have shown that there are more species than previously thought and there is still much to learn about some of these newly-minted turtles.

Photo © Mickle

ABOVE They may be small, but musk turtles are not shy about self-defense.

OPPOSITE The Intermediate Musk Turtle was only discovered in 2017, making it one of North America's newest turtles.

BELOW The small plastron of an Eastern Musk Turtle (**BOTTOM**), and the large, double-hinged plastron of an Eastern Mud Turtle (**BELOW**).

EASTERN MUSK TURTLE
Sternotherus odoratus

The Eastern Musk Turtle — formerly and affectionately called the Stinkpot — is widespread, but its secretive habits and small size make it difficult to see. It may be easiest to spot one from a canoe or kayak in clear water, as the turtle walks along the bottom of a shallow, weedy wetland.

Photo © Matt Brady

IDENTIFICATION Carapace is highly domed, narrow and smooth. Color can be nearly black, brown, gray or greenish. Plastron is very small and may be yellowish, brown or black. Head is relatively large with a pointed snout. Head and legs are dark. Two bright, pale yellow or whitish lines run along the face, above and below the eye. Carapace length up to 15 cm (6 in).

ALSO KNOWN AS Common Musk Turtle, Stinkpot Turtle.

SIMILAR SPECIES Compare to all other musk and mud turtles in its range.

RANGE Widespread in eastern North America. In the north from Maine west through central Ontario, and in the south from the tip of Florida west to central Texas.

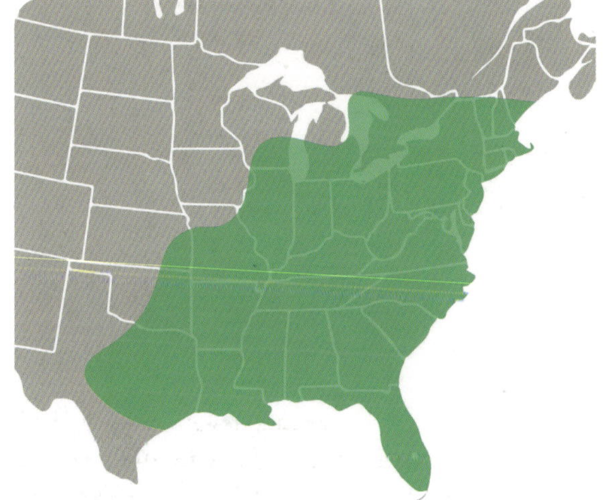

HABITAT Shallow areas in lakes, rivers and wetlands with soft bottoms, lots of vegetation and minimal current. Often basks by floating near the surface, hidden under lily pads or other vegetation. Occasionally climbs onto fallen trees or other debris.

DIET Invertebrates like snails, mussels, clams, crayfish, and insects, but also eats fish, fish eggs, amphibians, carrion, algae and plant material.

REPRODUCTION Lays up to nine eggs in a shallow nest, or simply under a log or other debris close to the water's edge. Up to two clutches per year.

CONSERVATION CANADA – Special Concern, USA – Not Listed, GLOBAL – Least Concern. Common in most parts of its range but declining in some areas. Destruction or alteration of wetlands may be its biggest threat.

RAZOR-BACKED MUSK TURTLE
Sternotherus carinatus

In a somewhat confusing group of turtles, the Razor-backed Musk Turtle's name and corresponding shell shape is a welcome bit of clarity. Unlike other ridged musk turtles, the carapace does not smooth over significantly with age, so the namesake feature is usually visible even on older adults.

IDENTIFICATION Carapace is highly domed and narrow, with a prominent central ridge that is present at all ages. Sides of the carapace slope away from the ridge, making the shell almost triangular in cross-section. Color is brownish or tan, with radiating dark lines that fade with age. Plastron is very small, and yellowish with variable dark smudging. Head is relatively large with a pointed snout. Head and legs are marked with dark spots. Carapace length up to 17 cm (7 in).

SIMILAR SPECIES Compare to Eastern and Stripe-necked Musk Turtle as well as Eastern and Yellow Mud Turtle which overlap its range.

RANGE The Gulf Coast of the United States from western Alabama to eastern Texas and as far north as central Arkansas and Oklahoma.

HABITAT Slow streams, rivers, and swamps with soft bottoms, lots of aquatic vegetation and woody debris. Usually does not leave the water, but occasionally will bask on exposed snags or vegetation.

DIET Invertebrates such as insects, snails, mussels, clams, and crayfish. Occasionally other small animals, carrion and plant material.

REPRODUCTION Lays up to seven eggs per clutch and up to three clutches per year. Little information exists on nest location, but likely similar to other musk turtles.

Photo © Evan Grimes

CONSERVATION USA – Not Listed, GLOBAL – Least Concern. Relatively common throughout its range. Pollution or alteration of habitat is a threat to this species.

TURTLES IN TREES

As a group, musk turtles spend most of their lives in the water, rarely leaving it for any reason other than nesting. Occasionally some species — especially the Razor-backed and Eastern Musk Turtle — make an exception to the rule and seek the sun on a dry perch.

Their small plastrons give musk turtles very flexible legs and they are surprisingly nimble climbers. When sunbathing, these turtles may climb onto the branch of a fallen tree and clamber as high as two meters above the water's surface.

Sitting high and dry comes with risk, as predators on land or in the air may make a meal of the small, exposed turtle. When a predator approaches they simply tuck into their shells and plummet to the safety of the water below. They have even been reported falling into canoes, unaware that the predator they perceived was actually a boat gliding beneath them.

Photo © Grover Brown

LOGGERHEAD MUSK TURTLE

Sternotherus minor

The Loggerhead Musk Turtle is well-named, as the large, bulbous head of the adult male looks as if it should tilt the turtle into a permanently face-down position. These turtles rarely leave the water and are seldom seen, so an encounter with one is a fortunate experience.

Photo © Jake Scott

IDENTIFICATION Carapace is highly domed and narrow, with three distinct ridges that may become smooth in older individuals. Color is brownish or grayish, with black streaks or spots that may fade with age. Plastron is small and yellowish with variable dark smudging. Head is large and blocky, especially in males, with a pointed snout. Head and legs are brown or gray and marked with dark spots. Carapace length up to 14 cm (5.5 in).

SIMILAR SPECIES Striped-necked and Intermediate Musk Turtles are extremely similar. A combination of range and neck markings is useful, but some individuals may not be identifiable in the field.

RANGE From central Florida north to most of Georgia and the eastern edge of Alabama.

HABITAT Springs, clear ponds, rivers, and streams with soft bottoms and vegetation or debris to provide shelter. Rarely leaves the water to bask.

DIET Invertebrates such as insects, mussels, snails, clams and crayfish. Occasionally other small animals, carrion and algae.

REPRODUCTION Lays up to five eggs in a nest dug in soft organic material near the water's edge. May lay up to five clutches per year.

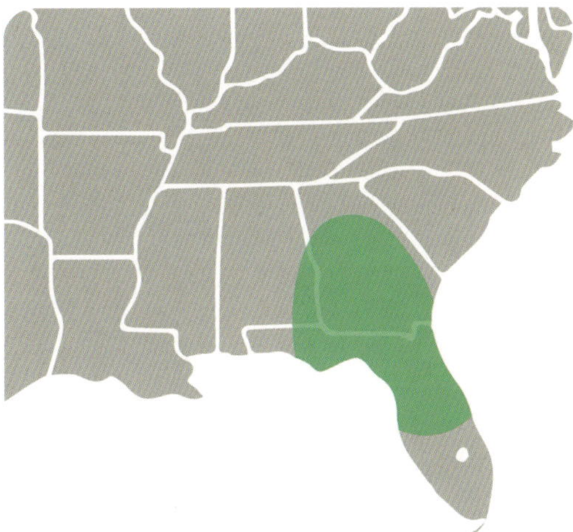

Photo © Kevin Hutcheson

CONSERVATION USA – Not Listed, GLOBAL – Least Concern. Relatively common throughout its range. Pollution or alteration of habitat is a threat to this species, as is collection for the pet trade.

A CONFUSING FAMILY TREE

The musk turtles may be one of the most confusing groups of North American turtles. Their relationships to each other have long been poorly understood. If you look through older turtle books you might find different names and arrangements of species and subspecies.

Most authorities have always recognized Loggerhead Musk Turtle as a species. Until very recently, it was divided into two subspecies: the Eastern Loggerhead Musk Turtle in the eastern part of the range, and the Stripe-necked Musk Turtle in the western part. Where the two subspecies met, they were believed to interbreed and create hybrids.

In the late 2010s, genetic research led by the University of Alabama produced some surprising results. Firstly, it showed that the Stripe-necked Musk Turtle was actually its own species, distinct from its cousin the Loggerhead. Incredibly, the evidence also suggested that the presumed hybrid turtles were not hybrids at all, but a

A turtle in an identity crisis.

Photo © Grover Brown

species all to themselves. This previously hidden species has been fittingly named Intermediate Musk Turtle.

Many authorities are adopting these changes to musk turtle taxonomy, but there is still inconsistency in how these turtles are described.

INTERMEDIATE MUSK TURTLE
Sternotherus intermedius

Until the late 2010s, this turtle was thought to be a hybrid between the Loggerhead and Stripe-necked Musk Turtles. Genetic research has shown it to be its own species, but little is yet known about its life history relative to its cousins.

Photo © Kevin Hutcheson

IDENTIFICATION Features are intermediate between Loggerhead and Stripe-necked Musk Turtle. Carapace has a central ridge and may have two additional ridges alongside it, though these may be small or absent and all ridges may smooth with age. Head is marked with spots and there is some striping on the sides of the head and neck. Range should be considered in identification, as this turtle only occurs in the Choctawhatchee and Escambia River basins. Carapace length up to 14 cm (5.5 in).

ALSO KNOWN AS Aliflora Musk Turtle

SIMILAR SPECIES Loggerhead and Stripe-necked Musk Turtles are extremely similar — combination of range and neck markings is useful, but some individuals may not be identifiable in the field.

Photo © Kevin Hutcheson

RANGE The western end of the Florida panhandle and southern Alabama.

HABITAT, DIET, & REPRODUCTION Because this species is newly discovered, little specific information exists on its biology as compared to the Loggerhead and Stripe-necked Musk Turtles. Most aspects are likely similar to these two closely related species.

CONSERVATION USA – Not Listed, GLOBAL – Not assessed. Little is known about the conservation of this new species, and future assessment will be needed to determine its status.

STRIPE-NECKED MUSK TURTLE
Sternotherus peltifer

The aptly named Stripe-necked Musk Turtle was once thought to be a subspecies of the Loggerhead Musk Turtle and has only recently been recognized as a species in its own right. There is much still to learn about this newly identified turtle species.

Photo © Kevin Hutcheson

IDENTIFICATION Carapace is highly domed and narrow, with a distinct central ridge that may flatten in older individuals. Color is brownish or tan with black streaks or spots that may fade with age. Plastron is small and yellowish with variable dark smudging. Head is relatively large and blocky with a pointed snout. Head and neck are marked with dark stripes on a tan or yellowish background. Carapace length up to 14 cm (5.5 in).

SIMILAR SPECIES Loggerhead and Intermediate Musk Turtles are extremely similar — combination of range and neck markings is useful, but some individuals may not be identifiable in the field.

RANGE From southern Alabama, Mississippi, and the eastern tip of Louisiana, north and east as far as western Virginia.

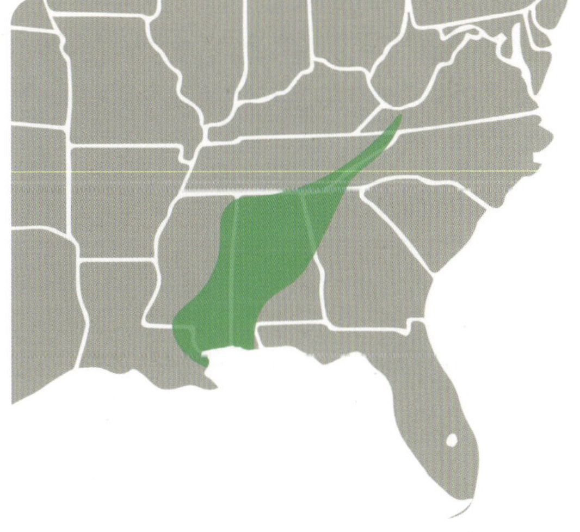

Photo © Kevin Hutcheson

HABITAT Almost exclusively rivers and streams with flowing water. Hides among snags and rocks along the bottom. Rarely leaves the water to bask.

DIET Invertebrates such as insects, snails, mussels, clams, and crayfish. Occasionally other small animals, carrion and algae.

REPRODUCTION Little specific information exists on reproduction in this newly-defined species. Likely similar to Loggerhead Musk Turtle.

CONSERVATION USA – Not Listed, GLOBAL – Not assessed. Little is known about the conservation of this new species, and future assessment will be needed to determine its status.

FLATTENED MUSK TURTLE
Sternotherus depressus

This little turtle is instantly recognizable. The typically high-domed shell of most musk turtles appears to have been squished flat in this species, almost as though the turtle has been stepped on. The Flattened Musk Turtle is critically endangered, so seeing one in the wild is very special indeed.

Photo © Kevin Hutcheson

IDENTIFICATION Carapace is narrow and flattened on the top with rounded sides. Young turtles have a slight ridge which flattens in adults. Color is light to dark brown, with dark spots that fade with age. Plastron is small and either pinkish or yellowish. Head is relatively large with a pointed snout. Head and legs are covered in a network of fine dark spots or lines. Carapace length up to 11 cm (4.5 in).

SIMILAR SPECIES Unique shape of carapace differentiates this species from all other musk turtles.

RANGE Endemic to just one small area in northern Alabama.

HABITAT Clear, shallow streams, creeks and rivers with sandy or rocky bottoms. Rarely leaves the water to bask. Spends much of its time hidden in crevices or under rocky cover.

DIET Snails, mussels, clams, insects and occasionally crayfish or carrion.

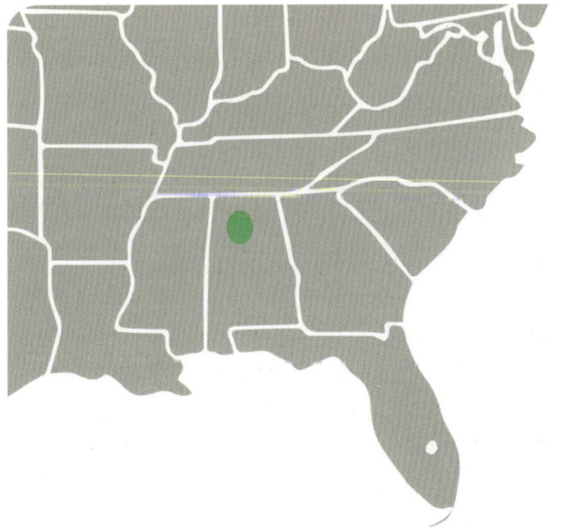

Photo © Kevin Hutcheson

REPRODUCTION Lays up to four eggs in a shallow nest dug in sandy soil not far from the water's edge. Up to two clutches per year.

CONSERVATION USA – Threatened, GLOBAL – Critically Endangered. Habitat degradation is the most significant threat, as this species needs clear, rocky streams to survive. Damming, pollution, erosion, and silt caused by mining, forestry, and agriculture are the leading causes of decline.

EASTERN MUD TURTLE
Kinosternon subrubrum

The Eastern Mud Turtle is the most widespread mud turtle in North America and the only one across much of its range. Seasonal migrations and even the search for food often bring this turtle onto land, so you're as likely to encounter one out of the water as in it.

IDENTIFICATION Carapace is highly-domed, narrow and smooth. Color ranges from light to dark brown or olive, with no obvious markings. Plastron is large and double-hinged, ranging from yellow to brownish and variably accented with dark markings. Head and neck are gray or brown with jagged pale streaks or mottling. Carapace length up to 12 cm (5 in).

ALSO KNOWN AS Common Mud Turtle

SIMILAR SPECIES Compare with all other mud and musk turtles that overlap its large range.

RANGE From New Jersey south to northern Florida, west to eastern Texas and Oklahoma.

HABITAT Shallow bodies of water with soft bottoms and abundant vegetation, including marshes, lagoons, swamps, ponds, ditches, and even wet fields. Has some tolerance for brackish water. Seldom basks but readily wanders over land.

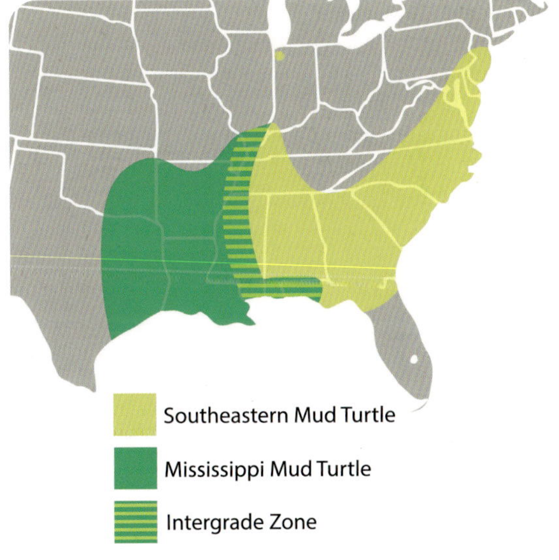

DIET Invertebrates like snails, mussels, clams, crayfish, and insects, but also amphibians, carrion, and plant material.

REPRODUCTION Lays up to six eggs in a nest dug in soft soil relatively near the water. Up to three clutches per year.

SUBSPECIES Divided into two very similar subspecies: Southeastern Mud Turtle (*K. s. subrubrum*) in the east and Mississippi Mud Turtle (*K. s. hippocrepis*) in the west. Mississippi Mud Turtle nearly always has two pale stripes on each side of the face, usually lacking in Southeastern. The two interbreed in a broad intergrade zone.

CONSERVATION USA – Not Listed, GLOBAL – Least Concern. Common throughout its range, except in the northernmost parts. Road mortality and loss of wetland habitat are its biggest threats.

FLORIDA MUD TURTLE
Kinosternon steindachneri

Until recently, the Florida Mud Turtle was considered a subspecies of the Eastern Mud Turtle. It is very similar to that species in many ways, but there is minimal overlap in range. The two species may hybridize where their ranges meet in northern Florida.

Photo © John Serrao

IDENTIFICATION Very similar to Eastern Mud Turtle. Carapace can be varying shades of brown with no obvious markings. Plastron is smaller than Eastern and the bridge that the connects the plastron to the carapace is narrower. Head and neck are gray or brown with no well-defined streaks or lines. Carapace length up to 12 cm (5 in).

SIMILAR SPECIES Striped Mud Turtle, Eastern Musk Turtle, and Loggerhead Musk Turtle share its range. Eastern Mud Turtle meets it at its northern boundary, where the two may hybridize.

RANGE Endemic to the state of Florida where it is relatively widespread but absent from the panhandle, the keys and an area east and south of Lake Okeechobee.

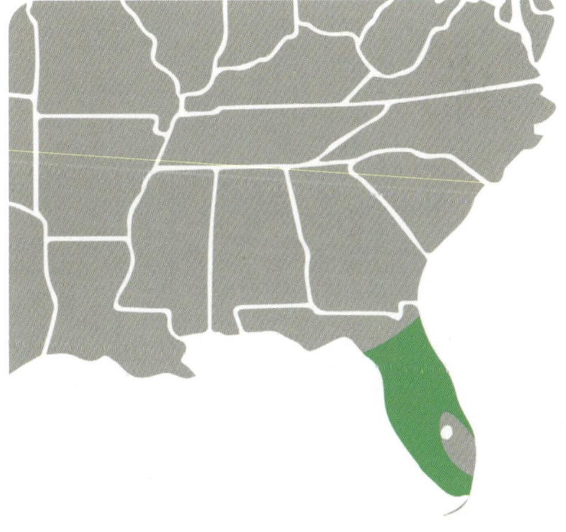

HABITAT Marshes, ponds, ditches and other small bodies of slow-moving water that are shallow, weedy and soft bottomed. Leaves the water less often than Eastern Mud Turtle.

DIET & REPRODUCTION Because this species is newly recognized, little specific information exists on its biology as compared to the Eastern Mud Turtle. Most aspects are likely similar to that closely related species.

CONSERVATION USA – Not Listed, GLOBAL – Not assessed. Reasonably common throughout its range. Further assessment is required for this newly described species.

STRIPED MUD TURTLE
Kinosternon baurii

Most Striped Mud Turtles are easily recognized by their namesake stripes, although the stripes may be hard to see on older adults. They are the most land-loving of all the mud turtles and will happily search for food out of the water, even picking insects out of cow poop.

IDENTIFICATION Carapace is highly-domed, narrow, and smooth. Color is usually dark brown to nearly black. Three pale stripes run along the top of the carapace, though they may fade with age. Plastron is large and double-hinged, ranging from yellowish to rusty brown. Head and neck are dark, and two yellow stripes mark each side of the face. Carapace length up to 12 cm (5 in).

ALSO KNOWN AS Three-striped Mud Turtle

SIMILAR SPECIES Compare to Eastern Mud Turtle, Florida Mud Turtle, Eastern Musk Turtle, and Loggerhead Musk Turtle which all occur within its range.

RANGE East coast states from Delaware south to the Florida Keys.

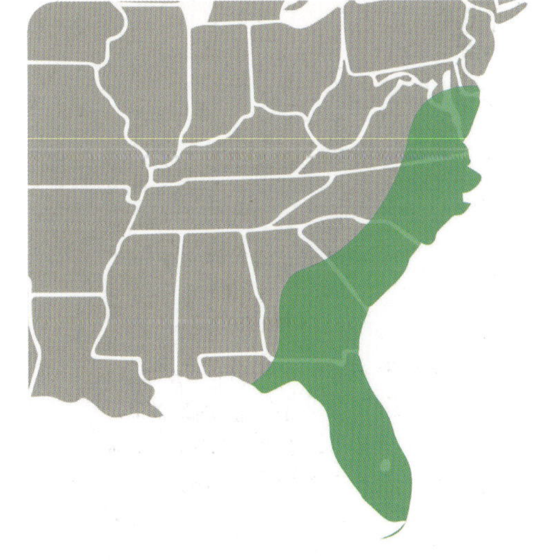

HABITAT Shallow bodies of water with soft bottoms and abundant vegetation, including marshes, swamps, ponds, ditches and canals. Some tolerance for brackish water. Frequently wanders on land.

DIET Invertebrates like snails, mussels, clams, crayfish and insects, but also amphibians, carrion and plant material.

REPRODUCTION Lays up to six eggs in a nest dug in sandy soil, sometimes several hundred meters from the water. Up to three clutches per year.

CONSERVATION USA – Not Listed, GLOBAL – Least Concern. Relatively common throughout its range.

ROUGH-FOOTED MUD TURTLE
Kinosternon hirtipes

North of Mexico, the Rough-footed Mud Turtle is only found in one small population in the Big Bend area of Texas. This is one of North America's most poorly studied turtles and only a little is known about its life history.

IDENTIFICATION Carapace is highly-domed, long, and narrow, and may have a faint ridge down the centre in all ages. Color ranges from light to dark brown or olive, with no obvious markings. Plastron is large and double-hinged, and yellowish with some black or brown markings. Head and neck are brown or greenish, and marked with dark spots or streaks. Carapace length up to 18 cm (7 in).

ALSO KNOWN AS Chihuahuan Mud Turtle

SIMILAR SPECIES In its range, only Yellow Mud Turtle is similar.

RANGE In the United States it occurs only in one small area near the Mexican border in western Texas.

HABITAT Naturally, rivers, streams and creeks in its otherwise arid environment, but now regularly uses human-made, spring-fed cattle tanks. Does not often leave the water.

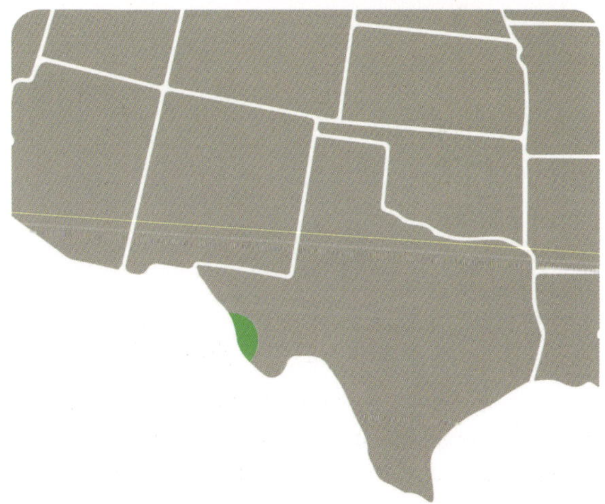

Photo © Miguel Gerardo Ochoa Tovar

DIET Limited information is available on the diet of this species but one study suggests it eats primarily insects and aquatic vegetation with occasional snails, seeds, fruits and algae.

REPRODUCTION Lays up to six eggs per clutch and potentially up to four clutches per year. Little information is available on the nest sites of this turtle.

SUBSPECIES Divided into six subspecies, but only one can be found north of Mexico: the Mexican Plateau or Big Bend Mud Turtle (*K. h. murrayi*).

CONSERVATION USA – Not Listed, GLOBAL – Least Concern. Not listed as at-risk but population is very small and restricted in range. Modification or destruction of stream habitat by agricultural practices and free-roaming cattle may threaten this species.

YELLOW MUD TURTLE
Kinosternon flavescens

A turtle of arid environments, the Yellow Mud Turtle will use almost any body of water available to it, migrating from one to another as they dry up. Its yellow throat is shared by its cousin the Arizona Mud Turtle but the two do not overlap in range, making identification relatively easy.

Photo © Bryan Box

IDENTIFICATION Carapace is domed, narrow, smooth and slightly flattened on top. Color is tan, olive-green or brown, with no obvious markings. Plastron is large and double-hinged, ranging from yellowish to brown. Head and neck are olive or brownish but throat and sides of neck are yellow. Carapace length up to 17 cm (7 in).

ALSO KNOWN AS Yellow-necked Mud Turtle

SIMILAR SPECIES In its range, compare to Sonoran Mud Turtle, Rough-footed Mud Turtle, Eastern Mud Turtle, Razor-backed Musk Turtle and Eastern Musk Turtle.

RANGE From eastern Texas north to Nebraska, and west to eastern New Mexico and the very southeastern tip of Arizona. An isolated population exists near the borders of Iowa, Illinois, and Missouri.

HABITAT Nearly any smaller body of water with a soft bottom, including ponds, marshes, rivers, cattle tanks, ditches and even flooded fields. Readily wanders over land.

DIET Invertebrates like snails, crustaceans, clams, worms and insects, but also amphibians, fish, carrion and plant material.

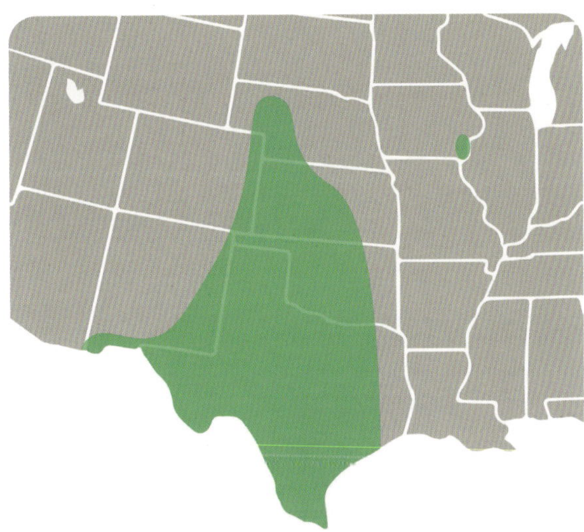

REPRODUCTION Lays up to nine eggs in a nest dug in sandy soil. Up to two clutches per year.

CONSERVATION USA - Not Listed, GLOBAL - Least Concern. Reasonably common in most parts of its range but declining in some states. Low water levels in these areas may be the cause of decline.

NEST GUARDIANS

Most turtles would not win any awards in the parenting department, as the vast majority simply lay their eggs and leave. The Yellow Mud Turtle, though, is a somewhat more doting parent.

These turtles begin their unusual nesting routine by first burying themselves in sandy soil. Once they are hidden well beneath the soil's surface, they dig a nest hole underneath themselves into which they lay their eggs. They then remain buried like this for anywhere from a few hours to over a month. Some may even overwinter in this position, emerging with their hatched young the following spring.

Exactly why the Yellow Mud Turtle does this is not totally clear. They may be protecting their eggs from predators like hognose snakes, rodents or raccoons. They may even be keeping the eggs moist by peeing on them. Whatever the case, their long stay underground with their eggs certainly sets them apart as standout turtle parents.

ARIZONA MUD TURTLE
Kinosternon arizonense

Once considered a subspecies of the Yellow Mud Turtle, the desert-dwelling Arizona Mud Turtle looks nearly identical to its cousin. The two do not overlap in range, so a map may be your most useful tool when identifying these species.

Photo © Paloma T. Montijo Tautimes

IDENTIFICATION Nearly identical to Yellow Mud Turtle with only small details of the carapace scutes differing. In the field, best identified by range. Carapace length up to 18 cm (7 in).

SIMILAR SPECIES In its range, only Sonoran Mud Turtle is similar.

RANGE In the United States, only occurs in a small area in southern Arizona near the Mexican border.

HABITAT Mostly temporary bodies of water, such as seasonally flooded pools and creeks, cattle tanks, and ditches. Frequently wanders over land, and often basks even in very hot weather.

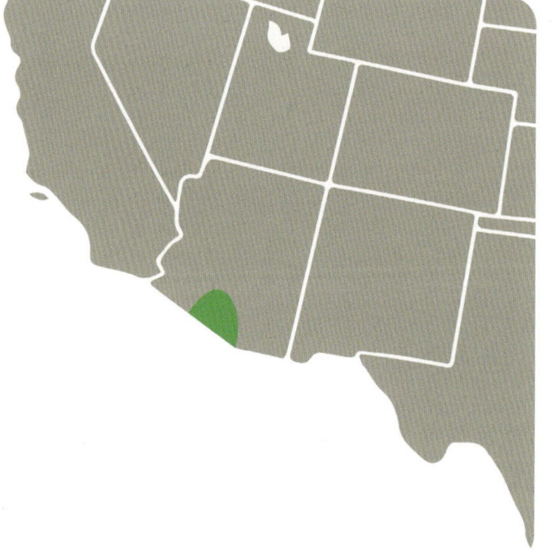

DIET Frogs, toads, and their tadpoles as well as insects and crustaceans.

REPRODUCTION Lays up to seven eggs per clutch and up to three clutches per year. Little information exists on nesting location.

CONSERVATION USA – Not Listed, GLOBAL – Least Concern. Reasonably common in its small range. Agricultural land and water use may affect this turtle, but the temporary impoundment of water for livestock may actually be beneficial.

Photo © Grover Brown

WAITING OUT THE DROUGHT

The life of an aquatic turtle in the desert has one very obvious challenge: the desert is very dry. Both the Arizona and Yellow Mud Turtles live in bodies of water that dry up seasonally. When the summer heat parches their habitats, drastic measures must be taken to survive.

As the water dwindles, these turtles simply bury themselves in the mud that remains in the pond or creek. Hidden from the glaring sun, their bodies enter a dormant state called aestivation. This is similar to hibernation, but occurs in the summer. This strategy allows the turtles to conserve energy and water until the rain falls once more.

Many of these turtles will aestivate from the summer of one year to the spring of the next. In a particularly dry year, they may not emerge at all, meaning that they simply skip an entire year and emerge almost two full years after they buried themselves.

Photo © Trevor B. Persons

SONORAN MUD TURTLE
Kinosternon sonoriense

Unlike our other desert-dwelling mud turtles, the Sonoran Mud Turtle prefers permanent bodies of water and is reluctant to wander over land. Heavy demand for water in the desert often puts these turtles in conflict with human activities.

IDENTIFICATION Carapace is highly-domed, narrow, and has three ridges that may smooth over with age. Color is olive or brown, with no obvious markings. Plastron is large and double-hinged, ranging from yellow to brownish. Head and neck strongly mottled with light and dark markings. Light markings may form stripes on the face. Carapace length up to 18 cm (7 in).

ALSO KNOWN AS Sonora Mud Turtle

SIMILAR SPECIES Yellow and Arizona Mud Turtles share its range.

RANGE Southwestern New Mexico west to southeastern California, though the population on the California/Arizona border may be extinct.

HABITAT Permanent bodies of water within its arid environment, including rivers, streams, ponds and cattle tanks. Prefers clear water with a rocky bottom. Rarely wanders over land.

Photo © Grover Brown

DIET Invertebrates like snails, crustaceans, and insects, but also amphibians, fish, carrion and plant material.

REPRODUCTION Lays up to 11 eggs in a nest dug in damp soil. Up to four clutches per year.

SUBSPECIES Divided into two subspecies: Desert Mud Turtle (*K. s. sonoriense*) and Sonoyta Mud Turtle (*K. s. longifemoralis*). The two are extremely similar but their ranges within the United States do not overlap.

CONSERVATION USA – Endangered (Sonoyta only), GLOBAL – Near Threatened. Diversion of water in the desert for irrigation and other human use is the primary threat to this turtle. Introduced bullfrogs and crayfish are also a threat, as they prey on young turtles. Because of its very small range in the United States, the Sonoyta subspecies is particularly at risk.

THE POND AND BOX TURTLES

- Painted Turtle
- Southern Painted Turtle
- Chicken Turtle
- Northern Red-bellied Cooter
- Florida Red-bellied Cooter
- Alabama Red-bellied Cooter
- River Cooter
- Peninsula Cooter
- Texas Cooter
- Rio Grande Cooter
- Pond Slider
- Big Bend Slider

THE MAP TURTLES
- Northern Map Turtle
- Barbour's Map Turtle
- Pascagoula Map Turtle
- Pearl River Map Turtle
- Escambia Map Turtle
- Alabama Map Turtle
- Black-knobbed Map Turtle
- Yellow-blotched Map Turtle
- Ringed Map Turtle
- Cagle's Map Turtle
- Texas Map Turtle
- False Map Turtle

- Ouachita Map Turtle
- Sabine Map Turtle

- Northwestern Pond Turtle
- Southwestern Pond Turtle
- Diamondback Terrapin
- Blanding's Turtle
- Spotted Turtle
- Bog Turtle
- Wood Turtle
- Common Box Turtle
- Ornate Box Turtle

ABOVE The cooters are among the most dedicated vegetarians of the North American turtles.

OPPOSITE The familiar Painted Turtle may be the archetypal pond turtle.

This is our largest family of turtles, with 35 species occurring north of Mexico. Despite their common name "pond turtles," members of this family make their homes in almost every body of water on the continent. Most of these turtles bask on logs, rocks and shorelines, so many of the most frequently seen turtles belong to this group.

This family is so large and varied that it is difficult to connect the species by any particular characteristic. Most have broad carapaces that are relatively low in profile, though the box turtles are a clear exception. All have large plastrons that provide ample protection below.

While the members of this family are diverse, there are three significant groups within the family that each comprise similar and closely related species:

- **THE COOTERS** (genus *Pseudemys*) — Some of our largest basking turtles, these seven species all grow to considerable size with deep, rounded shells. Recent genetic studies have

separated several species from each other but identification in the field can be difficult to near impossible for some individuals where ranges overlap. Unusual among the pond turtles, the cooters are primarily herbivores, eating very little animal matter as adults.

- **THE MAP TURTLES** (genus *Graptemys*) — These 14 species are extremely similar and very confusing. Research continues to elucidate new species within the group. Most have small, restricted ranges which can be significant aids to identification, but these turtles can be very difficult to differentiate when they co-occur. See page 118 for more information.
- **THE BOX TURTLES** (genus *Terrapene*) — This group contains two species and several subspecies, although their relationships are somewhat unclear and future study may split them apart. Unusual within the family, box turtles have high-domed carapaces and double-hinged plastrons which close completely into a protective box. This adaptation protects the box turtles as they wander on land, as this group is almost entirely terrestrial.

SUN WORSHIPPERS

Basking in the sun is a favorite pastime of nearly all turtles in this family and that is good news for the turtle-watcher. Resting on a log or shoreline, these turtles are conspicuous and easily observed. They are also vulnerable to predators, so most are quick to dash back into the water at the first sign of danger.

Turtles are endothermic, meaning that they cannot produce their own body heat. Basking allows them to raise their body temperature, which facilitates digestion and other metabolic processes. Some turtles, like snapping turtles and musk turtles, prefer to do their basking at the surface of the water, but there are advantages to the high-and-dry approach.

When a pond turtle basks on a log, it may completely dry out. This is not a problem for the turtle, but it is inhospitable to leeches and other parasites. It may even help prevent or treat fungal and bacterial infections, keeping the turtle fit and healthy. Basking out of the water seems to be worth the risk for these turtles, who are confident that their quick retreat will keep them out of harm's way.

OPPOSITE There's nothing like catching some rays on a summer day.

PAINTED TURTLE
Chrysemys picta

Easily the most naturally widespread turtle in North America, the Painted Turtle is also likely the most familiar. Within their range they inhabit almost any body of water and are easily identified by their colorfully marked necks, legs and shells.

IDENTIFICATION Carapace is relatively flat in profile and very smooth. Color is olive-green to almost black. Plastron is yellow with red and dark markings that vary among the subspecies. Head and legs are dark. Face is marked with pale stripes and spots. Neck, legs, and underside of carapace are beautifully marked with red or orange. Carapace length up to 25 cm (10 in).

SIMILAR SPECIES In its range, most similar to Northern Red-bellied Cooter, River Cooter and Pond Slider. At the southern edge of its range, compare to Chicken Turtle and Southern Painted Turtle.

RANGE Widespread across North America. In southern Canada from Nova Scotia to Vancouver Island, and in the United States as far south as Georgia and west to Oregon. Isolated populations occur in New Mexico and some have been introduced to California.

HABITAT Almost any body of slow-moving water with a soft bottom and abundant vegetation, including lakes, ponds, rivers and marshes. Very frequently basks on logs, rocks or shorelines.

DIET Insects, crustaceans, fish, carrion, plants and algae.

REPRODUCTION Lays up to 25 eggs in damp soil, usually not far from the water. Up to five clutches per year, but fewer in the northern parts of its range.

CONSERVATION CANADA – Special Concern, USA – Not Listed, GLOBAL – Least Concern. Common across most of its range. One Canadian population on the Pacific coast is designated Threatened. Road mortality and habitat loss or alteration have caused declines in many areas.

The Painted Turtle is divided into three subspecies.

Western Painted Turtle has a dark, patterned center on the plastron which is surrounded by red or orange that extends to the edges. A pattern of fine white lines may be present on the carapace.

Eastern Painted Turtle has an unmarked plastron. The scutes of the carapace form even rows across the back (unique among North American turtles).

Midland Painted Turtle has a dark center on the plastron which varies in prominence.

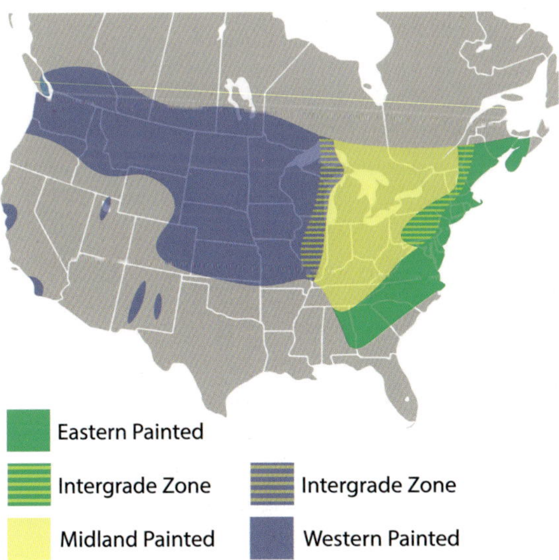

A VERY OFFICIAL TURTLE

What do elementary school students and state governments have in common? An affection for the Painted Turtle, apparently.

In 1994, students at the Cornwall School (now Bingham Memorial School) in Cornwall, Vermont suggested that the Painted Turtle should become the state's official reptile. The state government passed a resolution ratifying the choice, citing the turtle's hardworking nature, cold hardiness, beauty, beneficial pest control and commonality as justification.

In 1995, fifth graders in the city of Niles, Michigan followed the lead of the Cornwall students and successfully lobbied to have the Painted Turtle designated as that state's official reptile. The Painted Turtle was also elected as the state reptile of Illinois via an internet vote in 2005. In Denver, fourth graders led a campaign to make it Colorado's state reptile in 2008.

The Painted Turtle narrowly lost a vote for the same position in New York in 2006, where the equally deserving Snapping Turtle emerged the victor. Nevertheless, with official recognition in four states, the Painted Turtle holds the distinction of being the most honored state reptile.

SOUTHERN PAINTED TURTLE
Chrysemys dorsalis

The Southern Painted Turtle was long considered a subspecies of the Painted Turtle, but it has recently been separated as its own species. A colorful stripe distinguishes it from its more northern cousin, though the two may hybridize where their ranges meet.

Photo © Carl Franklin – TexasTurtles.org

IDENTIFICATION Very similar to Painted Turtle, but most have a prominent red, orange, or yellow stripe down the center of the carapace. Plastron is yellow and unmarked or with a faint dark patch in the center. Carapace length up to 15 cm (6 in).

SIMILAR SPECIES Shares its range with River Cooter, Alabama Red-bellied Cooter, Chicken Turtle and Pond Slider. At northern edge of its range, compare to Painted Turtle.

RANGE Gulf Coast from Alabama to Louisiana, north as far as southern Missouri.

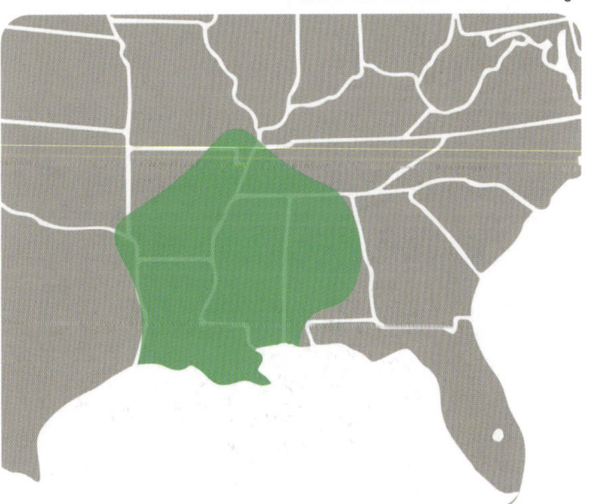

Photo © Carl Franklin – TexasTurtles.org

HABITAT Shallow bodies of water with soft bottoms and ample vegetation, including swamps, rivers, lakes, ponds and ditches. Basks frequently.

DIET Insects, snails, crayfish, algae and plant material. Occasionally fish and carrion.

REPRODUCTION Lays up to 20 eggs in sandy soil. Up to two clutches per year.

CONSERVATION USA – Not Listed, GLOBAL – Not Assessed. Common across most of its range but some declines have been observed. Road mortality and habitat loss or alteration are the biggest threats.

CHICKEN TURTLE
Deirochelys reticularia

The unfortunately named Chicken Turtle is apparently named for the taste of its meat, but it has plenty of other redeeming qualities. It may look round and clunky, but it is an active hunter and can quickly extend its long neck to snap up unsuspecting prey.

Photo © Peter Taylor

Photo © Michael Gallo/@ wildgallo

IDENTIFICATION Carapace is slightly rough and relatively flat in profile. It is often somewhat egg-shaped, being narrower at the front than the back. Color is dark greenish or brown. There is a network of yellow lines which may be prominent or quite faint. Plastron is yellow or orange and may have a dark marking in the center. Head and legs are dark with pale yellow lines. Front legs have a particularly broad yellow stripe. Upper part of hind legs has narrow, vertical lines. Neck is very long when extended. Carapace length up to 26 cm (10 in).

SIMILAR SPECIES Compare to all cooters, sliders and painted turtles in its range.

RANGE From North Carolina south to the tip of Florida, west through the Gulf Coast states to eastern Texas and spiking as far north as southern Missouri.

HABITAT Shallow bodies of still water with soft bottoms and abundant vegetation, including marshes, swamps, ponds, lakes, and ditches. Frequently wanders over land and will burrow into soil in forested areas to wait out dry weather conditions.

DIET Insects, crustaceans, snails, and tadpoles. Occasionally other invertebrates, fish, carrion and plant material.

Photo © Jedda Levy

Photo © Livan Escudero

Photo © Andrew Gluesenkamp

Photo © Andrew Heaton

Eastern Chicken Turtle is the largest subspecies and has an olive-brown carapace with a yellow outer margin.

Photo © Julian Grudens

Florida Chicken Turtle has the boldest yellow pattern on the carapace and a bright yellow or orange plastron.

Photo © Ethan Hollender

Western Chicken Turtle is the smallest subspecies. The plastron has dark markings along the edges of the scutes.

REPRODUCTION Lays up to 20 eggs in a nest dug in almost any soil, usually not far from the water. Up to four clutches per year. Unusually, it nests in the fall and winter and the hatchlings appear in the spring.

SUBSPECIES Divided into three subspecies: Eastern Chicken Turtle (*D. r. reticularia*), Florida Chicken Turtle (*D. r. chrysea*) and Western Chicken Turtle (*D. r. miaria*).

CONSERVATION USA – Not Listed, GLOBAL – Not Listed. Relatively common in most of its range, though some declines have occurred. Historically harvested for food, but now protected or regulated in most areas. Habitat loss to development or agriculture may be its primary threat.

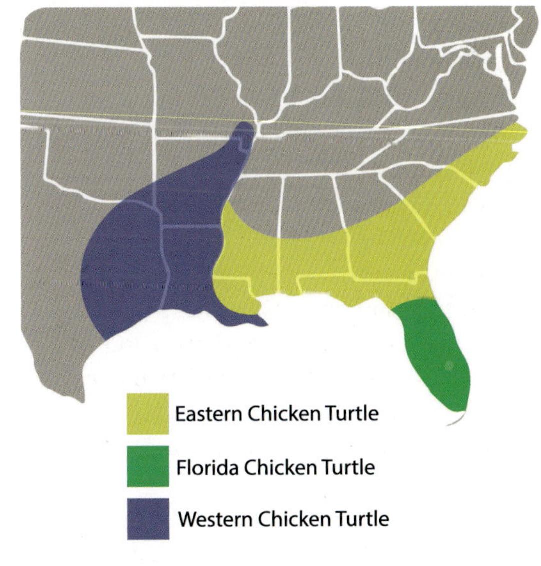

NORTHERN RED-BELLIED COOTER

Pseudemys rubriventris

In most of its range, the Northern Red-bellied Cooter is the largest basking turtle. Its large adult size, red carapace markings and bright red plastron set it apart from all other turtles in its habitat, making it an unusually easy cooter species to identify.

Photo © Randy L. Roberts

IDENTIFICATION Carapace is slightly rough and relatively flat in profile. Rear margin is slightly serrated. Color is brown, olive or black, with three reddish streaks that run from the center line to the margin on each side. Plastron is yellow with variable dark blotches and red around the edges that may be visible from above. Head and legs are dark with bold yellow or whitish lines. Some individuals become entirely black with age. Carapace length up to 40 cm (16 in).

ALSO KNOWN AS Plymouth Redbelly Turtle, American Redbelly Turtle.

SIMILAR SPECIES Other red-bellied cooters do not share its range. Compare to River Cooter, Pond Slider and Painted Turtle.

RANGE From Massachusetts south to North Carolina and as far west as West Virginia.

HABITAT Lakes, ponds, swamps and slow-moving rivers with soft bottoms and abundant vegetation. Basks very frequently on logs, rocks, vegetation or shoreline.

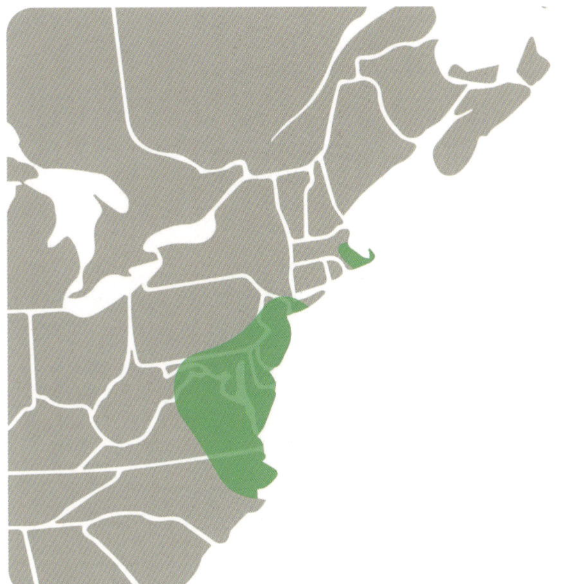

DIET Primarily plant matter as adults but young individuals will also eat insects and other invertebrates.

REPRODUCTION Lays up to 30 eggs in a nest in sandy soil. Up to two clutches per year.

CONSERVATION USA – Endangered, GLOBAL – Near Threatened. Only the Massachusetts population is federally endangered in the United States, while the remaining population is not listed. Habitat loss to development and agriculture, road mortality and nest predators are all risks to this species.

TURTLES DARK AND MYSTERIOUS

We all begin to look a little different as we age, but many cooters and sliders change so drastically that they become barely recognizable. These turtles may develop a condition called melanism and turn completely black as they get older. Melanism often obscures all other colors or markings on the turtles' skin.

If you encounter an entirely black turtle, determining what species you are looking at may take a little detective work. Cooters and sliders are the most likely to become melanistic, so checking which of those species live where you are is a good first step. Remember that the Pond Slider is introduced all over North America (and is often melanistic), so it is almost always a possibility.

Once you have narrowed the list down, you will need to look for other clues, such as any remaining markings on the turtle, as well as its overall shape and size. In areas where several potentially melanistic species occur together, the correct identification of some of these turtles may just remain a mystery.

Melanistic Pond Slider

Photo © Evan Grimes

FLORIDA RED-BELLIED COOTER
Pseudemys nelsoni

The big, deep-shelled Florida Red-bellied Cooter is a prominent feature of Florida wetlands and can be seen basking on any available surface in almost any body of water. It often lounges alongside the American Alligator, with which it has a tense and complicated relationship.

IDENTIFICATION Carapace is slightly rough, and moderately domed in profile. Rear margin is slightly serrated. Color is dark brown or black, with three reddish streaks that run from the center line to the margin on each side. Plastron ranges from yellow to orange or reddish, with variable dark markings towards the center. Head and legs are dark with bold yellow or whitish lines. Carapace length up to 38 cm (15 in).

ALSO KNOWN AS Florida Redbelly Turtle

SIMILAR SPECIES Other red-bellied cooters do not share its range. Compare to Peninsula Cooter, River Cooter and Pond Slider.

RANGE Southeastern Georgia and nearly all of Florida, including the Keys and parts of the panhandle.

HABITAT Almost any body of freshwater including lakes, ponds, rivers, swamps, marshes and canals. Somewhat tolerant of brackish water. Frequently basks.

DIET Primarily plant material as adults. Young turtles may also eat aquatic invertebrates and fish.

REPRODUCTION Lays up to 30 eggs in a nest dug in loose soil near the water. Up to six clutches per year.

CONSERVATION USA – Not Listed, GLOBAL – Least Concern. Common throughout its range.

A COMPLEX RELATIONSHIP

Adult cooters are hefty turtles and most have little to worry about as there are few predators that can contend with their thick shells. In the southeast, however, there is a creature that has just the right tools to deal with a heavily armored turtle: the American Alligator.

With formidable teeth, immense bite force and plenty of patience, the American Alligator is an awesome turtle hunter. The alligator is likely the reason that the cooters who live here have more highly domed carapaces than their northern cousins. The tall shell shape of the Florida Red-bellied Cooter makes it more resistant to crushing by the alligator's jaws than many flatter turtles.

There are limitations to this defense, though, and large alligators can crack even the toughest turtles. You might think that a cooter would avoid alligators at all costs, but there is another level to their relationship. The Florida Red-bellied Cooter sometimes digs its nest directly into the nest mound of an American Alligator. Like most turtles, the cooter leaves its eggs unguarded. The alligator, though, is fiercely protective of its nest, and in protecting its own eggs from predators it protects the turtle's, too. Whether the turtle knows this or is simply attracted to the soft soil for nesting, it could scarcely choose a better defender of its eggs.

ALABAMA RED-BELLIED COOTER
Pseudemys alabamensis

Found only in the deltas of a few river systems in southern Alabama and Mississippi, the endangered Alabama Red-bellied Cooter is unusually rare in a group of mostly common turtles. Protection of its habitat is crucial to the survival of this highly localized species.

Photo © Lauren McLaurin

IDENTIFICATION Carapace is slightly rough and moderately domed in profile. Rear margin is slightly serrated. Color is dark brown or greenish with three pale or reddish streaks that run from the center line to the margin on each side. Plastron ranges from yellow to orange or reddish, with variable dark markings towards the center. Head and legs are dark with bold yellow or whitish lines. Carapace length up to 38 cm (15 in).

ALSO KNOWN AS Alabama Redbelly Turtle

SIMILAR SPECIES Other red-bellied cooters do not share its range. Compare to Pond Slider and Southern Painted Turtle.

RANGE Restricted to southwestern Alabama and southeastern Mississippi.

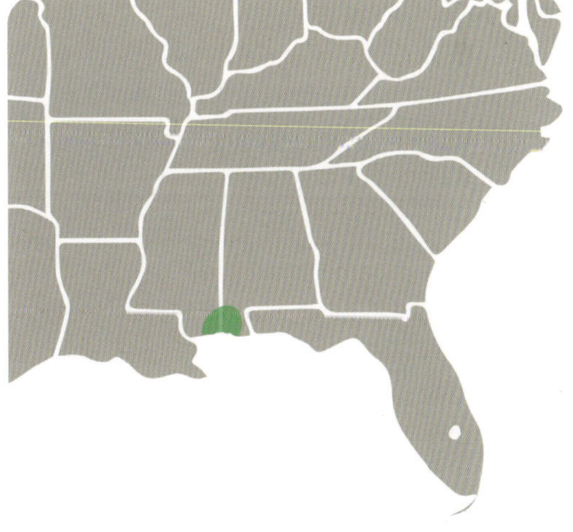

HABITAT Shallow areas in rivers, channels, and bays with soft bottoms and abundant vegetation. Some tolerance for brackish water. Frequently basks.

DIET Primarily plant material as adults. Young turtles may also eat aquatic invertebrates.

REPRODUCTION Lays up to 20 eggs in a nest dug in sandy soil relatively near the water. Up to three clutches per year.

CONSERVATION USA – Endangered, GLOBAL – Endangered. Found in only a very small range, where loss and alteration of vegetated riverine habitat is its biggest threat. Road mortality is also a significant risk for these turtles.

RIVER COOTER
Pseudemys concinna

The widespread River Cooter is the subject of some taxonomic confusion. Several other cooter species have previously been considered its subspecies, and at least two of its current subspecies are sometimes considered separate species.

Photo © Evan Grimes

IDENTIFICATION Carapace is slightly rough, relatively flat or moderately domed in profile and flared at the edges. Rear margin is slightly serrated. Color is greenish, brown, or black with a concentric pattern of orange or yellow lines that may fade with age. Dark, donut-shaped markings adorn the underside of the carapace. Plastron is yellow with variable dark markings. Head and legs are dark with many yellow lines. May become completely black with age. Carapace length up to 40 cm (16 in).

SIMILAR SPECIES Other cooters, sliders, and painted turtles.

RANGE From New York south to northern Florida and as far west as eastern Texas, Oklahoma, and Kansas. Several introduced populations occur on the west coast.

HABITAT Mostly rivers and lakes, but occasionally ponds, marshes, and other wetlands. Prefers rocky bottoms and abundant vegetation. Frequently basks, but rarely wanders far from the water.

DIET Primarily plant materials, but also insects, crayfish, tadpoles, fish, and other small animals.

REPRODUCTION Lays up to 30 eggs in loose or sandy soil quite near the water. Up to three clutches per year.

SUBSPECIES Divided into three subspecies: Eastern River Cooter (*P. c. concinna*), Coastal Plain or Florida River Cooter (*P. c. floridana*) and Suwannee River Cooter (*P. c. suwanniensis*). The status of these subspecies is debated. Both Coastal Plain and Suwannee River Cooter are sometimes considered to be separate

Photo © Evan Grimes

species. All three are very similar and field identification extremely difficult where their ranges meet.

CONSERVATION USA – Not Listed, GLOBAL – Least Concern. Relatively common throughout most of its range but considered at-risk in some states. Road mortality and habitat alteration are the biggest threats.

TURTLE TAXONOMY TROUBLES

Taxonomy is the science of sorting and classifying living things. Taxonomists use all kinds of evidence to do this, including what an animal looks like, the structure of its bones or organs, where it lives, its genetic makeup, its behavior and even the fossil record. Classification can be confusing at the best of times and new evidence is always emerging, so taxonomy is always changing!

The River Cooter is an excellent example of just how challenging taxonomy can be. This book lists three subspecies, two of which are sometimes considered to be separate species. Older texts often list two additional subspecies — The Missouri and Hieroglyphic River Cooters — which most newer sources no longer recognize. The Peninsula, Texas, and Rio Grande Cooters have each been considered subspecies of the River Cooter at times, too. Making matters worse, different organizations or authorities may use different information when deciding which species or subspecies to recognize.

So, in all this chaos and confusion, how do we know what is right? The simple answer, unfortunately, is that there is no simple answer. Taxonomy is fluid and there is no single authority that we can turn to. All we can do is use the best information we have at a given moment and remember that things may get a little confusing.

A River Cooter. Probably.

PENINSULA COOTER
Pseudemys peninsularis

The Peninsula Cooter illustrates just how befuddling cooters can be, as it is sometimes considered a subspecies of the Florida River Cooter when that turtle is not itself considered a subspecies of the River Cooter.

IDENTIFICATION Carapace is slightly rough and moderately domed in profile. Rear margin is slightly serrated. Color is greenish, brown, or black with pattern of yellow lines that run parallel to each other. Pattern may fade with age. Plastron is yellow and usually unmarked. Head and legs are dark with many yellow lines. Carapace length up to 40 cm (16 in).

ALSO KNOWN AS Peninsular Cooter

SIMILAR SPECIES In its range, compare to River Cooter, Florida Red-bellied Cooter, and Pond Slider.

RANGE Extreme southern Georgia and most of Florida, including the Keys but excluding the western part of the panhandle.

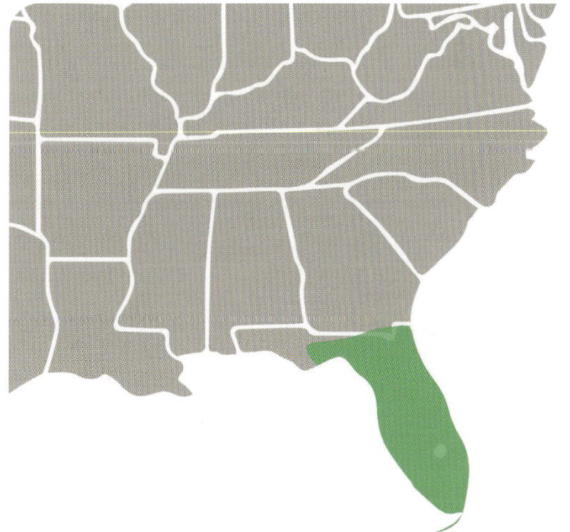

Photo © Grover Brown

HABITAT Any slow-moving body of water with abundant vegetation and a soft bottom, including lakes, swamps, marshes and canals. Basks frequently.

DIET Primarily plant material as adults. Young turtles may also eat aquatic invertebrates and fish.

REPRODUCTION Lays up to 20 eggs in a nest in sandy soil not far from the water. Up to three clutches per year.

CONSERVATION USA – Not Listed, GLOBAL – Least Concern. Relatively common throughout its range.

TEXAS COOTER
Pseudemys texana

Likely the most common big, basking turtle throughout most of its range, the well-named Texas Cooter is endemic to the Lone Star State. This turtle is a river specialist so watch for it in flowing bodies of water.

Photo © Annika Lindqvist

IDENTIFICATION Carapace is slightly rough and relatively flat in profile. Rear margin is slightly serrated. Color is dark brown or greenish with pattern of concentric yellow lines that fades with age. Plastron is pale yellow with dark lines along the scute edges. Head and legs are dark with many yellow lines. A short, vertical line is usually present just behind the eyes. Carapace length up to 35 cm (14 in).

ALSO KNOWN AS Texas River Cooter

SIMILAR SPECIES In its range, compare to River Cooter and Pond Slider.

RANGE Endemic to river basins of eastern and central Texas.

Photo © Rebecca Louise Goddammit

HABITAT Primarily rivers, but sometimes also lakes, canals, cattle tanks, and ditches. Prefers abundant vegetation. Basks frequently.

DIET Primarily plant material as adults. Young turtles may also eat a variety of aquatic invertebrates.

REPRODUCTION Lays up to 20 eggs in a nest not far from the water. Up to three clutches per year.

CONSERVATION USA – Not Listed, GLOBAL – Least Concern. Relatively common throughout its range.

RIO GRANDE COOTER
Pseudemys gorzugi

Native only to the Rio Grande and its tributaries, the Rio Grande Cooter is arguably the prettiest turtle of this group. Alteration of rivers has led to its disappearance from many places in which it once lived and it now inhabits only the least-altered stretches of the watershed.

Photo © Roberto Gonzalez

IDENTIFICATION Carapace is slightly rough and relatively flat in profile. Rear margin is slightly serrated. Color is olive green or brown with a pattern of concentric black, yellow and/or reddish lines that may fade with age. Edges and underside of carapace may have heavy reddish markings. Plastron is pale yellow with dark lines along the scute edges. Head is dark with pale yellow lines and, often, larger blotches behind the eyes. Legs are striped with yellow, orange or reddish color. Carapace length up to 38 cm (15 in).

SIMILAR SPECIES In the Rio Grande watershed it is only likely to be confused with Big Bend Slider and introduced Pond Slider.

RANGE The Rio Grande basin along much of the border between Texas and Mexico, reaching north into southeastern New Mexico.

HABITAT Primarily slow-moving rivers but sometimes ponds and lakes. Prefers soft bottoms and abundant vegetation. Basks frequently.

Photo © Celestyn Brożek

DIET Primarily plant material, but more omnivorous than other cooters and will eat a variety of invertebrates and other small animals.

REPRODUCTION Limited information exists on the reproduction of this turtle. It may lay up to 20 eggs per clutch, and up to two clutches per year.

CONSERVATION USA – Not Listed, GLOBAL – Near Threatened. Has disappeared from much of its historic range in the Rio Grande watershed. Pollution from oilfields, damming, and intensive water use are likely the primary causes of the decline.

POND SLIDER
Trachemys scripta

Even if you've never seen one in the wild, you've likely encountered the Pond Slider in your neighborhood pet store. These charismatic turtles have traveled the world in the pet trade and are now an invasive species in many places outside their native range.

IDENTIFICATION Carapace is slightly rough and relatively flat in profile. Rear margin is slightly serrated. Color is greenish or brown with pattern of yellow markings that typically includes bold yellow lines that run from the midline to the margin. Pattern often fades with age. Plastron is yellow with variable dark markings. Head and legs are dark with pale yellow lines. Head pattern varies by subspecies (see opposite). Older individuals may become completely black. Carapace length up to 30 cm (12 in).

SIMILAR SPECIES Shares its range with almost all other pond turtles. Especially compare with cooters, Big Bend Slider and painted turtles.

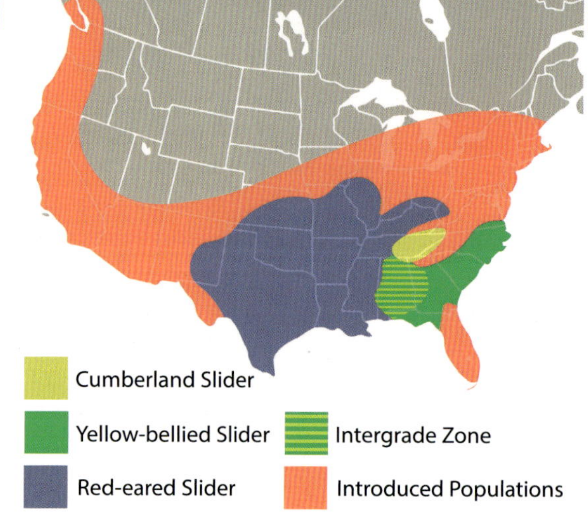

Red-eared Slider

Yellow-bellied Slider

Cumberland Slider

RANGE Naturally occurs from Virginia south to northern Florida, north as far as Illinois and west as far as New Mexico and western Texas. Now widely introduced where it has been released from captivity and can occur almost anywhere in North America. Introduced populations are primarily of the Red-eared subspecies.

HABITAT Almost any body of fresh water, including ponds, lakes, rivers, canals, marshes, swamps and ditches. Prefers water that is unmoving or slow moving with a soft bottom and abundant vegetation. Basks frequently.

DIET Widely varied, including plant material, algae, various aquatic invertebrates, fish, tadpoles, other small animals and carrion.

REPRODUCTION Lays up to 30 eggs in a nest dug in various soil types, sometimes far from the water. Up to five clutches per year.

SUBSPECIES Divided into three subspecies: Red-eared Slider (*T. s. elegans*), Yellow-bellied Slider (*T. s. scripta*), and Cumberland Slider (*T. s. troostii*). Red-eared Slider has a conspicuous red stripe behind the eye. Yellow-bellied Slider has a broad yellow blotch in same location. Cumberland Slider has fewer and broader yellow stripes on face and neck. Facial markings may fade with age in all subspecies, making identification difficult.

CONSERVATION USA – Not Listed, GLOBAL – Least Concern. Common throughout its range and introduced across nearly all of North America.

THIS TURTLE GETS AROUND

Pond Sliders are popular as pets and have been both harvested from the wild and bred in captivity for sale in the pet trade since the early 1900s. Often sold as babies in pet stores they quickly grow into large and unmanageable adults. Many well-intentioned pet owners, realizing they cannot provide for their new pets, have released these turtles into their local ponds or wetlands.

Fortunately for the turtle, but unfortunately for local ecosystems, the Pond Slider is a survivor. These hardy creatures are adaptable to a wide array of habitats and climates. They have thrived across North America and have established populations in Europe, Asia and even Australia. Today, you can see a Pond Slider almost anywhere in the world.

Pond Sliders are so hardy that they often outcompete local turtles for food, basking sites and other resources. They may also introduce diseases. These enthusiastic invaders can be catastrophic where they are introduced, so releasing captive turtles into the wild should be avoided at all costs.

The ubiquitous presence of the Pond Slider means that you should consider it as a possibility whenever you are watching and identifying turtles anywhere in North America.

BIG BEND SLIDER
Trachemys gaigeae

The Big Bend Slider is similar to the Pond Slider, but unlike its cosmopolitan cousin it is relatively rare. This turtle is native only to parts of the Rio Grande River and its tributaries, where alteration of its riverine habitat puts it at risk.

Photo © Missy McAllister-Kerr

IDENTIFICATION Carapace is slightly rough and relatively flat in profile. A weak central ridge may be present. Rear margin is slightly serrated. Color is greenish or brown with a pattern of orange lines and dark marks that may form eye-like spots. The pattern often fades with age. Plastron is yellow or orange with a central dark marking. Head and legs are dark with many yellow lines. Two oval-shaped red or orange spots (one large and one small) bordered with black appear behind the eye. Older individuals may become completely black. Carapace length up to 30 cm (12 in).

ALSO KNOWN AS Mexican Plateau Slider

SIMILAR SPECIES The Pond Slider is most similar, but details of head and carapace should differentiate between the two. Also compare to Rio Grande Cooter and Painted Turtle.

RANGE The Rio Grande basin from western Texas north into New Mexico.

HABITAT Rivers and nearby ponds, canals and cattle tanks, usually with abundant vegetation.

Photo © Jody Shugart

DIET Primarily plant matter and algae as adults. Young individuals will also eat insects, other invertebrates and carrion.

REPRODUCTION Lays up to 30 eggs in a nest dug in soft or sandy soil. Up to two clutches per year.

CONSERVATION USA – Not Listed, **GLOBAL** – **Vulnerable.** Intensive water use, pollution, river damming and diversion in the Rio Grande watershed are all significant risks to this turtle. Hybridization with the introduced Pond Slider may also be a threat to this species.

THE MAP TURTLES

ABOVE The "lips" of a map turtle can often be seen from a distance.

The following 14 species of map turtle form a close subgroup (or genus) within the Pond and Box Turtle family. They are closely related and often extremely similar in appearance and behavior, so some general information about map turtles is provided here. Many of these species are endemic to single river systems along the Gulf Coast of the United States, while just a few are more widespread.

Because of their similarities, identifying map turtles can be challenging. Fortunately, their restricted ranges are very helpful in narrowing the possibilities. Knowing what river system you are looking at can be your most important tool in identifying these tricky turtles.

APPEARANCE & BEHAVIOR

Map turtles have a rough-textured carapace that is relatively low in profile but has a central ridge. The ridge is often accented with spikes, which wear down in older adults. The rear margin is often flared and always serrated. The color is usually olive-green or brown with a variable pattern of pale lines. This network of lines may resemble those on a topographic map, giving rise to the name of this group. The plastron is typically pale yellow and may be unmarked or have some dark markings.

The head and legs are generally dark and covered in pale yellow or greenish lines. The head may have spots or blotches which can be helpful in identification. The edges of the beak are pale, giving map turtles the appearance of having lips. The webbed hind feet are especially large, and map turtles are powerful swimmers.

Map turtles may have the most extreme sexual dimorphism of any pond turtle, with adult females being much larger than adult males. In the Barbour's Map Turtle, for example, the females may reach a carapace length of 32 cm (13 in) with the much smaller males rarely exceeding 12 cm (5 in). The two can be so different in appearance that they may seem like totally different species.

Map turtles are adept swimmers in deep water and rarely stray far from their aquatic homes. They love to bask but are very wary, plunging into the water when they perceive any threat, even at great distance.

REPRODUCTION

Some of the map turtles are among the most poorly studied of all North American turtles. There is little specific information on the reproductive biology of several species. Most lay fewer than 20 eggs in nests dug in sandy soil, seldom very far from the water. Some lay only a single clutch per year, while others can lay as many as four.

CONSERVATION

Map turtles are very sensitive to poor water quality and changes in their environments. Many species, especially those that live in rivers along the American Gulf Coast, are threatened or endangered. The main threats to these turtles are pollution, water degradation, damming and modification of rivers caused by mining, forestry and agricultural practices. Collection for the pet trade, and even target shooting for sport, also pose major risks to these species.

ABOVE The intricate lines that give this group its name are often most visible on young individuals.

NORTHERN MAP TURTLE
Graptemys geographica

The widespread Northern Map Turtle is the only map turtle in significant portions of its range. Its rough shell does not shine in the sun like the smoother shells of other basking turtles. It is further set apart by its central ridge and serrated rear margin.

IDENTIFICATION Carapace is greenish, brown or gray, with a complex network of pale, dark-bordered lines that fade with age. Central ridge may be slightly spiked in small individuals but projections almost never present in large turtles. A small, yellowish spot appears behind each eye. Carapace length up to 27 cm (11 in).

ALSO KNOWN AS Common Map Turtle

SIMILAR SPECIES Compare to False Map Turtle and Ouachita Map Turtle, which occur in parts of its range.

Photo © Kyle Horner

RANGE Widespread, from Quebec south to New York and west as far as Minnesota, Kansas and Oklahoma.

HABITAT Large bodies of water such as rivers, lakes and bays, occasionally smaller creeks or lake-shore marshes.

DIET Primarily hard-shelled prey such as mussels, snails and crayfish. Also aquatic insects and carrion.

CONSERVATION CANADA – Special Concern, USA – Not Listed, GLOBAL – Least Concern.

Native only to rivers in southeastern Alabama, southwestern Georgia and the Florida panhandle, the Barbour's Map Turtle doesn't usually occur alongside any of its close cousins.

Photo © Grover Brown

IDENTIFICATION Carapace is greenish to dark olive or gray, with C-shaped yellow markings that fade with age. Prominent central ridge has black-tipped spikes which flatten in older adults. A pale blotch behind each eye connects to another blotch on top of the head. Heads of females may be very large and blocky. Carapace length up to 32 cm (13 in).

SIMILAR SPECIES The only map turtle in most of its range. Escambia Map Turtle is the only species that may overlap.

RANGE From the Gulf Coast in the central Florida panhandle, north into western Georgia and eastern Alabama.

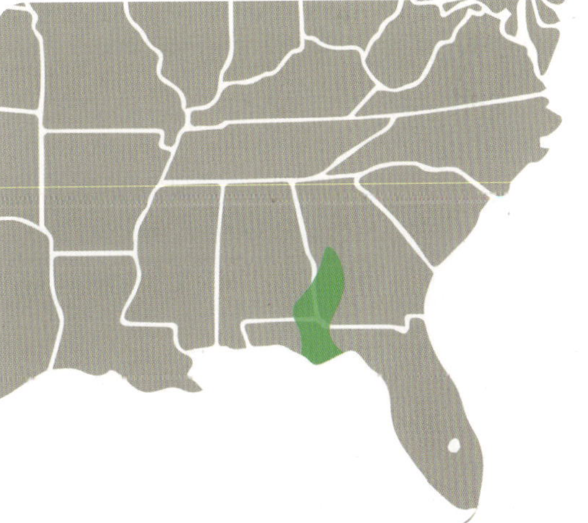

Photo © Joe Girgente [@ joes_outdoor_adventures]

HABITAT Deep, fast-flowing rivers and streams with rocky bottoms.

DIET Mussels, clams, snails, and insects.

CONSERVATION USA – Proposed Threatened, GLOBAL – Vulnerable. A listing of Threatened under the Endangered Species Act of the United States has been proposed for this species.

PASCAGOULA MAP TURTLE
Graptemys gibbonsi

The Pascagoula Map Turtle is conveniently named for the only river system in which it occurs. The Yellow-blotched Map Turtle shares its range almost exactly, so watch for these two on the Pascagoula River.

IDENTIFICATION Carapace is olive green or brown, and each marginal scute has a wide, pale yellow or orange bar. Prominent central ridge has a black stripe, and several spikes which flatten in older adults. A pale blotch behind each eye connects to another blotch on top of the head. Heads of females may be very large and blocky. Carapace length up to 30 cm (12 in).

SIMILAR SPECIES Shares its range with Yellow-blotched Map Turtle. Also compare to nearby Ringed, Pearl River, Black-knobbed and Alabama Map Turtles.

RANGE Endemic to the Pascagoula River system in southeastern Mississippi.

Photo © Grover Brown

HABITAT Deep, fast-flowing rivers and streams with sand or gravel bottoms.

DIET Clams, sponges, insects, and some plant material.

CONSERVATION USA – Proposed Threatened, GLOBAL – Endangered. A listing of Threatened under the Endangered Species Act of the United States has been proposed for this species.

PEARL RIVER MAP TURTLE
Graptemys pearlensis

The Pearl River Map Turtle is another species named for the only river in which it lives. The Ringed Map Turtle shares its range almost exactly.

Photo © Evan Grimes

IDENTIFICATION Nearly identical to Pascagoula Map Turtle, of which it was once considered a subspecies, but does not overlap in range. Pale bars on the marginal scutes tend to be narrower. Carapace length up to 30 cm (12 in).

SIMILAR SPECIES Shares its range with the Ringed Map Turtle and may overlap with Ouachita and False Map Turtle.

RANGE Endemic to the Pearl River system in southern Mississippi and eastern Louisiana.

Photo © Evan Grimes

HABITAT Deep, fast-flowing rivers and streams with sand and gravel bottoms.

DIET Clams, mussels, insects, and occasionally snails and fish.

CONSERVATION USA – Proposed Threatened, GLOBAL – Endangered. A listing of Threatened under the Endangered Species Act of the United States has been proposed for this species.

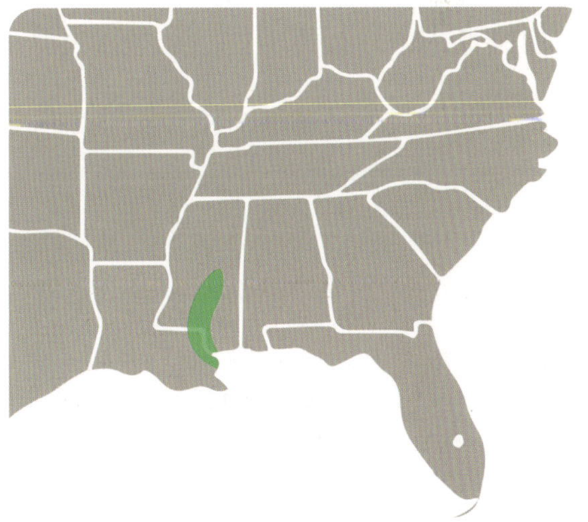

ESCAMBIA MAP TURTLE
Graptemys ernsti

The Escambia Map Turtle is, for the most part, the only map turtle that occurs within its range. It is native only to rivers that drain into Escambia Bay.

Photo © Jonathan Mays

IDENTIFICATION Carapace is olive green or brown. Each marginal scute has a broad, pale, C-shaped marking. Similar markings may be present on the large scutes of the carapace. Prominent central ridge has a broken black stripe and several spikes which flatten in older adults. A pale blotch behind each eye does not connect to the blotch on top of the head. Heads of females may be very large and blocky. Carapace length up to 28 cm (11 in).

SIMILAR SPECIES Compare to neighboring Black-knobbed, Barbour's, and Alabama Map Turtles.

RANGE Southern Alabama and the western end of the Florida panhandle.

HABITAT Deep, fast-flowing rivers and streams with rocky bottoms.

Photo © Grover Brown

DIET Mussels, clams, snails, crayfish, and insects.

CONSERVATION USA – Proposed Threatened, GLOBAL – Near Threatened. A listing of Threatened under the Endangered Species Act of the United States has been proposed for this species.

Photo © Owen Kinney

IDENTIFICATION Carapace is olive green or brown with C-shaped yellow markings that fade with age. Prominent central ridge has a black stripe and several spikes which flatten in older adults. A pale blotch behind each eye connects to a large pale area on top of the head. Carapace length up to 30 cm (12 in).

SIMILAR SPECIES Shares its range with Black-knobbed Map Turtle. At edges of its range, may coincide with Escambia, False, Northern, Ouachita, Pascagoula and Yellow-blotched Map Turtles.

RANGE Much of Alabama and neighboring areas in Mississippi, Tennessee, Georgia and possibly the tip of the Florida panhandle.

Photo © Grover Brown

HABITAT Fast-flowing rivers and streams with sandy, muddy or rocky bottoms.

DIET Snails, mussels, clams, sponges, insects and some plant material.

CONSERVATION USA – Proposed Threatened, GLOBAL – Near Threatened. A listing of Threatened under the Endangered Species Act of the United States has been proposed for this species.

BLACK-KNOBBED MAP TURTLE
Graptemys nigrinoda

The Black-knobbed Map Turtle is usually easy to identify since it bears a clear, identifying feature that you can see at a distance. If only they were all so simple.

Photo © Toby J. Hibbitts

IDENTIFICATION Carapace is olive green or brown, with hollow, circular, yellow markings that are bordered in black. Rear margin is strongly serrated. Marginal scutes have C-shaped markings. Pattern often fades with age. Very prominent central ridge has several black, knobby projections that flatten somewhat in older adults. Often has a pale crescent behind each eye that opens rearward. Carapace length up to 20 cm (8 in).

ALSO KNOWN AS Black-knobbed Sawback

SIMILAR SPECIES Shares its range with Alabama Map Turtle. At edges of its range, may coincide with Escambia, False, Northern, Ouachita, Pascagoula, and Yellow-blotched Map Turtles.

RANGE Rivers throughout Alabama and eastern Mississippi.

Photo © Jeff Garner

HABITAT Fast-flowing rivers and streams with sandy or clay bottoms.

DIET Mussels, snails, sponges, other aquatic invertebrates, and some plant material.

CONSERVATION USA – Not Listed, GLOBAL – Least Concern.

YELLOW-BLOTCHED MAP TURTLE
Graptemys flavimaculata

The yellow blotches of the Yellow-blotched Map Turtle are a dead giveaway when they're visible. This species lives only in the Pascagoula River system in southern Mississippi.

Photo © Grover Brown

IDENTIFICATION Carapace is olive green or brown, usually with large yellow patches on each large scute. Rear margin is strongly serrated. Marginal scutes have C-shaped markings. Pattern often fades with age. Very prominent central ridge has several black-tipped spikes that flatten somewhat in older adults. A pale spot appears behind each eye. Carapace length up to 20 cm (8 in).

ALSO KNOWN AS Yellow-blotched Sawback

SIMILAR SPECIES Shares its range with Pascagoula Map Turtle. Also compare to nearby Ringed, Pearl River, Black-knobbed, and Alabama Map Turtles.

RANGE Endemic to the Pascagoula River system in southeastern Mississippi.

HABITAT Fast-flowing rivers and streams with sand or gravel bottoms.

DIET Primarily sponges, clams, mussels, insects, and some plant material.

CONSERVATION USA – Threatened, GLOBAL – Vulnerable.

RINGED MAP TURTLE
Graptemys oculifera

Perhaps the prettiest map turtle when its pattern shows clearly, the Ringed Map Turtle lives only in the Pearl River, alongside its cousin that bears the Pearl River name.

Photo © Aaron Luke Madsen

IDENTIFICATION Carapace is olive green or brown, with hollow, circular, yellow markings that are bordered in black. Rear margin is strongly serrated. Marginal scutes have C-shaped markings. Pattern often fades with age. Very prominent central ridge has several black-tipped spikes that flatten somewhat in older adults. A pale spot appears behind each eye. Carapace length up to 22 cm (9 in).

ALSO KNOWN AS Ringed Sawback

SIMILAR SPECIES Shares its range with the Pearl River Map Turtle, and may overlap with Ouachita, and False Map Turtle.

RANGE Endemic to the Pearl River system in southern Mississippi and eastern Louisiana.

HABITAT Deep, fast-flowing rivers and streams with sand and gravel bottoms.

DIET Poorly studied, but seemingly insects and some plant material.

CONSERVATION USA – Threatened, GLOBAL – Vulnerable.

CAGLE'S MAP TURTLE
Graptemys caglei

The Cagle's Map Turtle lives only in three river basins in southern Texas. Only the False Map Turtle occurs alongside it, limiting identification confusion.

Photo © Viviana Ricardez – TexasTurtles.org

IDENTIFICATION Carapace is olive green or brown, with many dark and light lines arranged in concentric rings. Pattern often fades with age. Central ridge has spikes when young, which may flatten completely in older adults. Top of the head is marked with a forward-opening, curved V-shape, the ends of which may form a crescent behind each eye. Carapace length up to 22 cm (9 in).

SIMILAR SPECIES Only False Map Turtle overlaps its range. Also compare to the nearby Texas Map Turtle.

RANGE Endemic to the Guadalupe, San Antonio and San Marcos River basins in southern Texas.

Photo © Carl Franklin - TexasTurtles.org

HABITAT Deep, fast-flowing rivers and streams with rocky, sandy or muddy bottoms.

DIET Snails, clams, mussels, crayfish and insects.

CONSERVATION USA – Not Listed,
GLOBAL – Endangered.

TEXAS MAP TURTLE
Graptemys versa

The Texas Map Turtle may live in Texas, but it confusingly calls the Colorado River (and associated water bodies) home.

Photo © Viviana Ricardez – TexasTurtles.org

IDENTIFICATION Carapace is olive green or brown with a network of fine, yellow lines that often fade with age. Central ridge has yellow-tipped spikes when young. The spikes may flatten completely in older adults. A short, orange or yellow line, often J-shaped, extends horizontally behind each eye. Carapace length up to 22 cm (9 in).

SIMILAR SPECIES Only False Map Turtle overlaps its range. Compare to the nearby Cagle's Map Turtle.

RANGE Endemic to the Colorado River drainage in southern and central Texas.

Photo © Jim & Lynne Weber

HABITAT Rivers with slow to moderate current and associated lakes and streams with rocky bottoms.

DIET Clams, mussels, snails, insects, sponges, and occasionally other invertebrates, algae, and plant material.

CONSERVATION USA – Not Listed,
GLOBAL – Least Concern.

FALSE MAP TURTLE
Graptemys pseudogeographica

There's nothing false about the confusingly named False Map Turtle. This unusually widespread map turtle is commonly kept as a pet and is found in many places outside of its native range where it has been released from captivity.

Photo © Luís Lourenço

IDENTIFICATION Carapace is olive green or brown with pale lines surrounding dark blotches, although this pattern may be very faint and fades with age. Central ridge has black-tipped spikes when young, which may flatten completely in older adults. Markings around the eyes differ between subspecies (see opposite). Carapace length up to 27 cm (11 in).

SIMILAR SPECIES Large range overlaps with many other map turtle species. The False Map has often been introduced outside of its native range.

RANGE From Ohio north and west as far as North Dakota and south to Mississippi, Louisiana and eastern Texas. Boundaries of the large intergrade zone between the subspecies are not entirely clear. Often occurs outside its native range where it has been released from the pet trade.

HABITAT Primarily large rivers and creeks, but occasionally lakes and ponds. Prefers a muddy bottom and some vegetation.

DIET Insects, snails, crayfish, mussels, carrion, and plant material.

Photo © Dustin Lynch, Arkansas Natural Heritage Commission

Mississippi Map Turtle head pattern.

Northern False Map Turtle head pattern. Photo © Michael Long

SUBSPECIES Divided into two subspecies: Northern False Map Turtle (*G. p. pseudogeographica*) and Mississippi Map Turtle (*G. p. kohnii*). Northern has a short, crosswise line behind each eye which may form an "L" shape. Mississippi has a line which forms a crescent, wrapping around the back of each eye.

CONSERVATION USA – Not Listed, GLOBAL – Least Concern.

OUACHITA MAP TURTLE

Graptemys ouachitensis

Like many of its cousins, the Ouachita Map Turtle inhabits a single river basin. That river basin just happens to be the mighty Mississippi, giving this turtle an unusually large range among its kin.

IDENTIFICATION Carapace is olive green or brown with a network of dark-bordered yellow lines surrounding dark blotches. This pattern may be very faint and fades with age. Central ridge has black-tipped spikes when young but which may flatten in older adults. A squarish pale blotch behind the eye often tapers to a narrow point or line on top of the head and may join with a small, oval-shaped spot under the eye. Carapace length up to 27 cm (11 in).

ALSO KNOWN AS Southern Map Turtle

SIMILAR SPECIES Large range overlaps with many other map turtle species, and Ouachita Map is sometimes introduced outside of its native range.

RANGE Roughly follows the Mississippi River basin north from the Louisiana coast, widening across Oklahoma, Arkansas and Tennessee, then

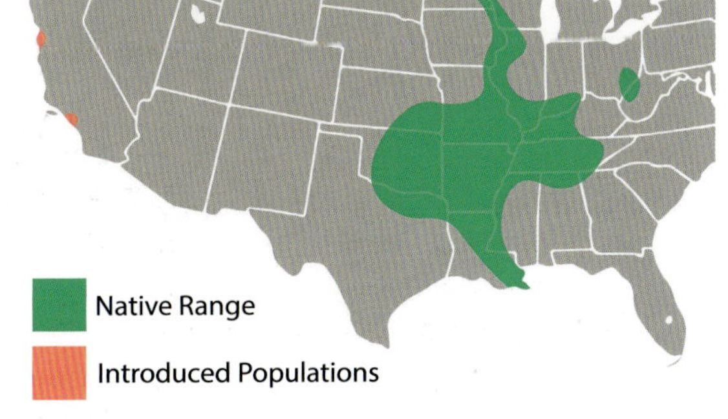

tapering northward towards the river's headwaters in Minnesota.

HABITAT Primarily large rivers and creeks, but occasionally lakes and ponds. Prefers a muddy bottom and some vegetation.

DIET Insects, snails, crayfish, mussels, carrion, and plant material.

CONSERVATION USA – Not Listed, GLOBAL – Least Concern.

SABINE MAP TURTLE
Graptemys sabinensis

Until recently, the Sabine Map Turtle was a subspecies of the Ouachita Map Turtle so minimal research has been done on it specifically. It is endemic to the Sabine River system of Louisiana and Texas.

Photo © Grover Brown

IDENTIFICATION Nearly identical to Ouachita Map Turtle with which it was once considered the same species. Pale spot behind the eye is usually round or oval and surrounded by concentric lines. Range should be a significant factor in identification. Possibly the smallest map turtle. Carapace length may not exceed 16 cm (6 in).

SIMILAR SPECIES False Map Turtle shares its range. Very similar to nearby Ouachita Map Turtle.

RANGE Endemic to the Sabine River basin in western Louisiana and eastern Texas.

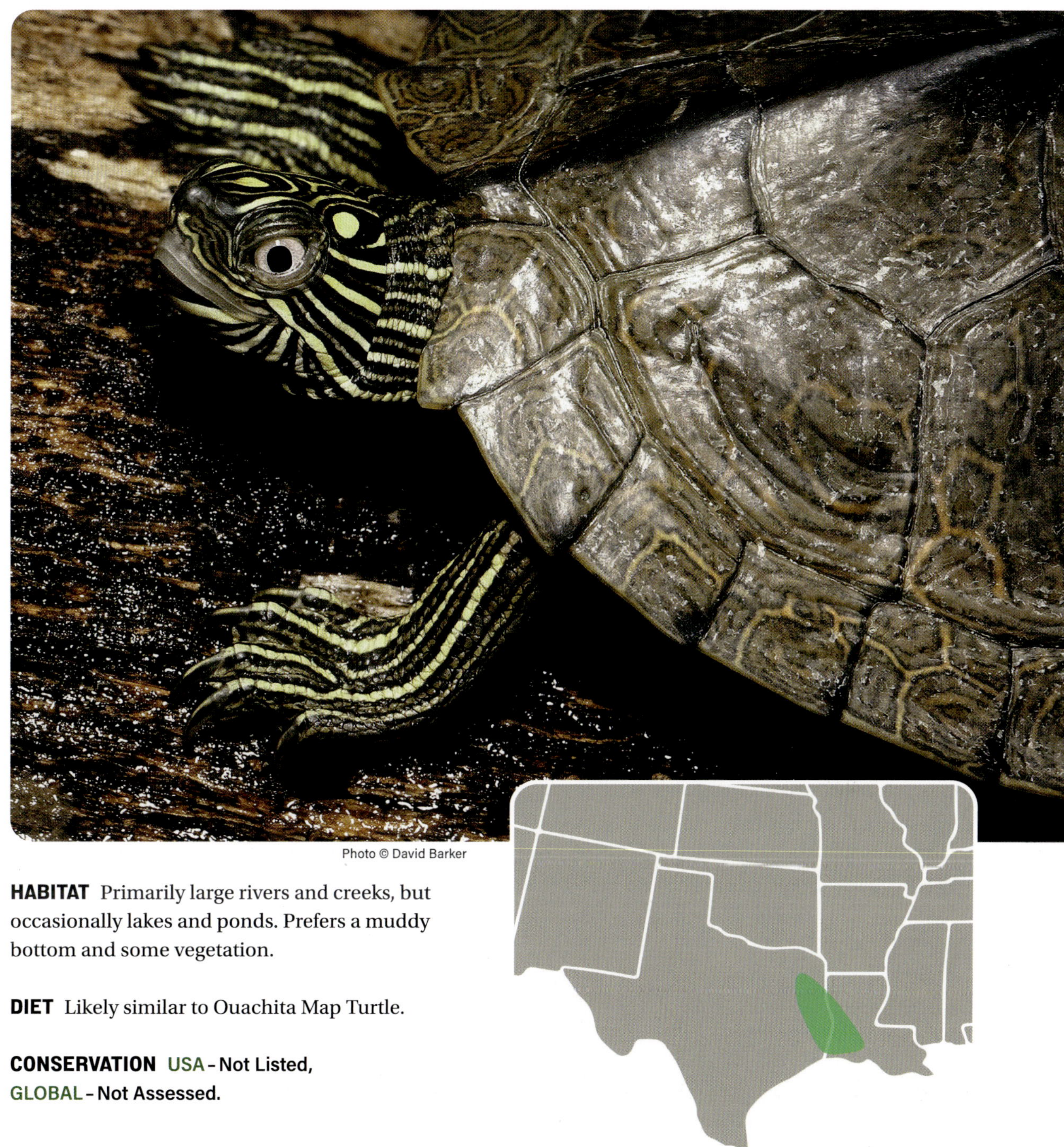

Photo © David Barker

HABITAT Primarily large rivers and creeks, but occasionally lakes and ponds. Prefers a muddy bottom and some vegetation.

DIET Likely similar to Ouachita Map Turtle.

CONSERVATION USA – Not Listed, GLOBAL – Not Assessed.

NORTHWESTERN POND TURTLE
Actinemys marmorata

One of few turtles that inhabit the west coast of North America, the Western Pond Turtle was recently split into two species: the Northwestern and Southwestern Pond Turtle. The two are very similar but don't overlap geographically except in the narrow band where their ranges meet.

IDENTIFICATION Carapace is smooth, broad and low in profile. Color is greenish, brown or nearly black, with a pattern of fine, lighter spots or lines that radiate outward from the center of each large scute. Pattern may fade with age or simply appear mottled. Plastron is yellow and may have a dark blotch on each scute. Head and legs have a dark base color but are usually heavily flecked with yellowish spots. Adult males have an unmarked, pale chin and throat. Carapace length up to 24 cm (9 in).

ALSO KNOWN AS (Northern) Pacific Pond Turtle

SIMILAR SPECIES Few pond turtles share the western range of this species. Range mostly separates the closely related Southwestern Pond Turtle. Compare to Painted Turtle and the introduced Pond Slider and River Cooter.

RANGE From Washington south to the San Francisco Bay area of California and continuing south from the Central Valley eastward. Extirpated from British Columbia and some parts of Washington where it once lived.

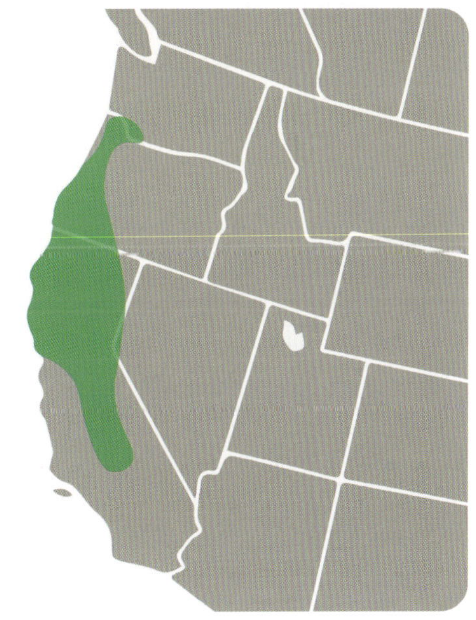

HABITAT Most bodies of fresh water with deep pools and aquatic vegetation, including lakes, ponds, rivers, streams, and marshes. Some tolerance for brackish water. Frequently basks and regularly wanders over land in forests or grasslands.

DIET Widely varied, including insects, crayfish, other invertebrates, amphibians, carrion and plant material.

REPRODUCTION Lays up to 13 eggs in a nest dug in sandy soil, often near water but sometimes hundreds of meters away. Up to two clutches per year.

CONSERVATION CANADA – Extirpated, USA – Not Listed, GLOBAL – Vulnerable. Historically present in southern British Columbia, Canada, but has disappeared from there and other areas in the northern parts of its range. Heavily impacted by extensive historical harvesting for food. Still threatened by habitat alteration, competition with invasive Pond Sliders and predation of young turtles by invasive Largemouth Bass and American Bullfrogs.

SOUTHWESTERN POND TURTLE
Actinemys pallida

Only recently considered a separate species from its similarly named cousin, the Southwestern Pond Turtle is one of just a few turtle species native to coastal California.

Photo © Zack Abbey

IDENTIFICATION Extremely similar to the Northwestern Pond Turtle, perhaps best identified by range. Where ranges meet, identification of individuals in the field may be impossible. One small difference in the scutes of the plastron (the presence or absence of an inguinal plate) that is often used to separate the species requires up-close examination and may be inconsistent. Carapace length up to 24 cm (9 in).

ALSO KNOWN AS (Southern) Pacific Pond Turtle

SIMILAR SPECIES Few pond turtles share the western range of this species. Range mostly separates the closely related Northwestern Pond Turtle. Compare to Painted Turtle and the introduced Pond Slider and River Cooter.

RANGE Coastal California south of the San Francisco Bay area and west of the Central Valley to the Mexican border.

HABITAT Most bodies of fresh water with deep pools and aquatic vegetation, including lakes, ponds, rivers, streams and marshes. Some

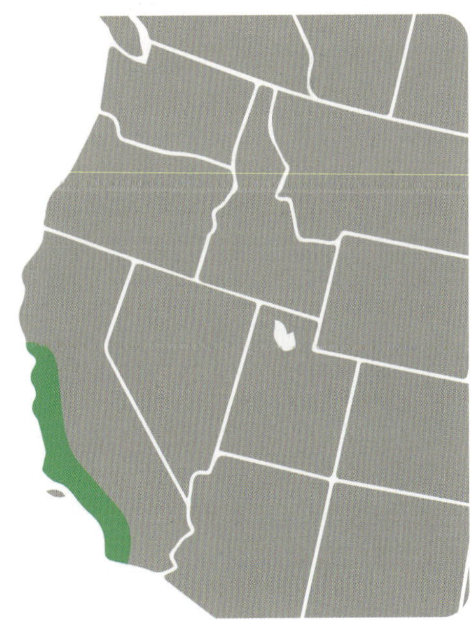

tolerance for brackish water. Frequently basks and regularly wanders over land in forests or grasslands.

DIET Widely varied, including insects, crayfish, other invertebrates, amphibians, carrion and plant material.

REPRODUCTION Lays up to 13 eggs in a nest dug in sandy soil, often near water but sometimes hundreds of meters away. Up to two clutches per year.

CONSERVATION USA – Not Listed, GLOBAL – Vulnerable. Heavily impacted by extensive historical harvesting for food. Still threatened by habitat alteration, extreme weather events on the California coast, competition with invasive Pond Sliders and predation of young turtles by invasive Largemouth Bass and American Bullfrogs.

DIAMONDBACK TERRAPIN
Malaclemys terrapin

Easily a contender for both the prettiest and strangest turtle in North America, the Diamondback Terrapin is like a pond turtle that dreams of being a sea turtle. The terrapin lives exclusively in brackish, coastal habitats that no other turtles call home, leaving the water only to bask or nest on sandy beaches.

IDENTIFICATION Carapace is low in profile, with concentric ridges and grooves on each large scute. There is a central ridge, which is weak in some subspecies and prominent in others, occasionally with spikes or knobs. Color ranges from pale gray or brown to almost black and concentric dark and light rings often trace the scute ridges. Plastron may be yellow, orange or gray and and there are variable dark markings. Head and legs are pale gray or brownish and boldly marked with black spots and squiggles. Beak is pale, giving the appearance of lips. Carapace length up to 23 cm (9 in).

ALSO KNOWN AS Diamond-backed Terrapin

SIMILAR SPECIES No other turtle regularly shares the terrapin's brackish or saltwater habitats. Its unique appearance makes it unmistakable.

RANGE Along the east coast of the United States from Massachusetts to nearly the southern tip of Texas.

HABITAT Coastal marshes, mangrove swamps, river mouths, lagoons and shallow bays along the seashore. Can tolerate both saltwater and freshwater, but primarily lives where the water is intermediate between the two (brackish). Frequently basks but rarely wanders far from the water.

DIET A variety of marine invertebrates, including clams, mussels, snails, shrimp, crabs, barnacles and worms. Also eats fish and occasionally carrion.

REPRODUCTION Lays up to 22 eggs in a nest dug in a sand dune close to the shore. Up to three clutches per year.

SUBSPECIES There are seven recognized subspecies: Northern (*M. t. terrapin*), Carolina (*M. t. centrata*), Florida East Coast (*M. t. tequesta*), Mangrove (*M. t. rhizophorarum*), Ornate (*M. t. macrospilota*), Mississippi (*M. t. pileata*), and Texas (*M. t. littoralis*). There is much variation among

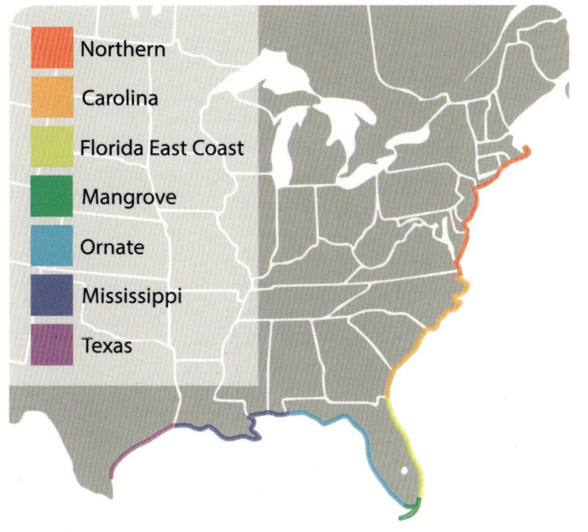

Photo © Evan Grimes

subspecies and they may interbreed, so geography is the easiest way to separate them and individuals where the ranges of two subspecies meet may not be identifiable.

CONSERVATION USA – Not Listed, GLOBAL – Vulnerable. Extensively hunted for food through much of the 1900s. Current threats include shoreline development or alteration, accidental drowning in crab traps, road mortality, and subsidized predators on their nesting beaches.

WATER, WATER EVERYWHERE...

As any mariner worth his salt will tell you, being surrounded by water does not guarantee you a drink. The Diamondback Terrapin knows this problem all too well, as the salty water it lives in is no good for drinking.

A turtle's kidneys, much like our own, are not able to deal with large quantities of salt. When confronted with excess salt, they try to excrete it in urine. This excretes water, too, which is why drinking seawater can, counterintuitively, make you dehydrated.

The terrapin, then, needs a way to drink water without also ingesting salt. It does this by slurping up the rainwater that forms as a thin film on surfaces and even on top of the seawater itself. If it rains hard, the terrapin will simply look up and open its mouth like a child catching a snowflake, drinking rainwater straight from the source.

Even with these clever behaviors, a terrapin is certain to ingest at least a little seawater. To deal with a buildup of salt in its body, this pond turtle uses the same solution as the fully marine sea turtles. A gland beside the eye, called a lachrymal gland, excretes a fluid more concentrated with salt than seawater. These salty tears keep the turtle's body at salt levels lower than the water around it, allowing it to survive in environments that most turtles cannot tolerate.

BLANDING'S TURTLE
Emydoidea blandingii

The ever-smiling Blanding's Turtle with its cheerful yellow throat is always a delight to see. Its high, smooth, helmet-like shell stands out at a distance and it often basks alongside the somewhat flatter Painted Turtle and rougher map turtles.

IDENTIFICATION Carapace is long, smooth and highly-domed. Color is black, dark gray or dark brown and heavily speckled with pale yellow flecks. Plastron is hinged at the front, allowing partial closure of the shell. Plastron color is yellow with a large, black patch on each scute. Head and legs are dark with some yellow flecking. Chin and underside of neck are entirely bright yellow. Neck is very long when extended. Carapace length up to 28 cm (11 in).

SIMILAR SPECIES Not easily confused with other species. Compare with Spotted Turtle and box turtles.

RANGE From the extreme western edge of Quebec through Ontario and the states that border the Great Lakes, as far west as Nebraska. Isolated populations occur in Nova Scotia, New York, Massachusetts, New Hampshire and Maine.

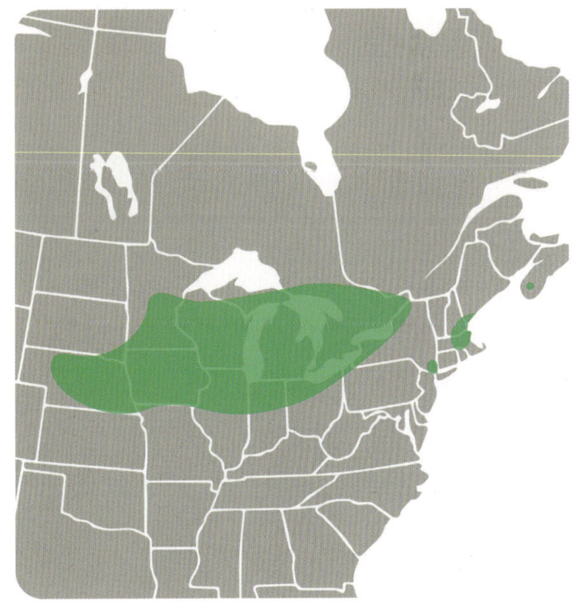

HABITAT Shallow bodies of fresh water with soft bottoms and lots of vegetation, including lakes, marshes, ponds and ditches. Regularly basks and readily wanders on land.

DIET Crayfish, insects, snails, fish, amphibians, carrion and some plant material.

REPRODUCTION Lays up to 25 eggs in a nest dug in loose or sandy soil, sometimes over a kilometer from the water. Rarely more than one clutch per year.

CONSERVATION CANADA – Endangered, USA – Not Listed, GLOBAL – Endangered. Road mortality is an especially significant threat to these turtles, particularly during their long migrations in search of nest sites. Loss of their wetland habitat also puts them at risk.

The diminutive Spotted Turtle, adorned with perfect polka-dots, is about as cute as a turtle can be. Incredibly, this tiny turtle may live for well over 100 years. Collection of the long-lived adults for the pet trade is devastating to populations, so locations of this rare turtle are often a closely guarded secret.

Photo © Evan Grimes

IDENTIFICATION Carapace is low in profile and smooth. Color is black or dark gray, with a varying number of round yellow spots. Plastron is orange or yellow with a large, black patch on each scute. In some individuals the black patch may grow to cover the entire scute. Head and legs are dark with variable yellow or orange spots and blotches. Males have brown eyes and a dark chin while females have orange eyes and a yellow or orange chin. Carapace length up to 13 cm (5 in).

SIMILAR SPECIES The perfect spots on the carapace make this turtle mostly unmistakable. Compare to Bog Turtle and Blanding's Turtle.

RANGE Ontario and the states surrounding the lower Great Lakes, eastward to Maine and south through all east coast states to northern Florida.

HABITAT Shallow bodies of slow-moving fresh water with soft bottoms and lots of vegetation,

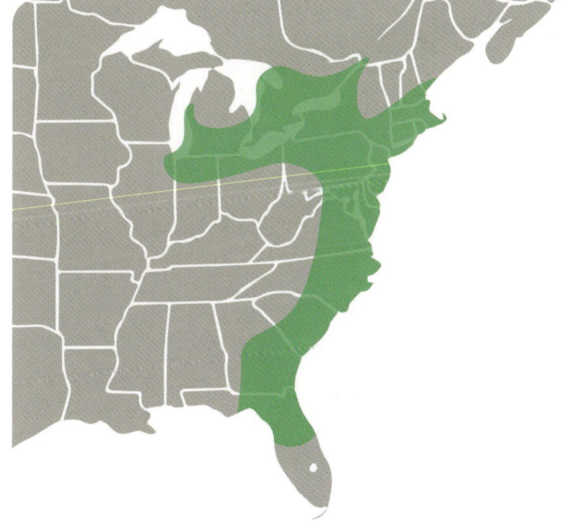

including swamps, fens, bogs, marshes and wet meadows. Sometimes basks and occasionally wanders on land near the water.

DIET Insects, leeches, snails, crayfish, small fish, amphibians and some plant material.

REPRODUCTION Lays up to seven eggs in a nest dug in loose or sandy soil, often near but sometimes hundreds of meters from the water. Up to three clutches per year. Multiple clutches are more likely in the southern part of its range.

CONSERVATION CANADA – Endangered, USA – Not Listed, GLOBAL – Endangered. Alteration and development of the shallow, weedy habitats that this turtle needs are probably the most significant threat. Illegal collection for the pet trade is also a big risk to this species.

BOG TURTLE
Glyptemys muhlenbergii

The adorable Bog Turtle is both the smallest turtle in North America and one of the rarest. Its rarity, combined with its small size and secretive habits, make it a turtle you are unlikely to come across, and an opportunity to see one in the wild would be very special indeed.

IDENTIFICATION Carapace is moderately domed with a weak central ridge. Each scute has concentric grooves and ridges, though some individuals may be fairly smooth. Color is black or dark brown. There may be paler lines radiating from the center of each scute. Plastron is yellow with a variable amount of dark marking. In some turtles, the entire plastron may be dark. Head and legs are dark, with a large yellow or orange blotch behind each eye. The blotch may be broken. Carapace length up to 11 cm (4.5 in).

SIMILAR SPECIES The mostly unmarked carapace and colorful head blotches make this turtle easily recognizable. Compare to Spotted Turtle and Wood Turtle.

RANGE Several disjunct populations ranging from New York and Massachusetts in the north to South Carolina and Georgia in the south.

HABITAT Shallow wetlands with muddy bottoms and lots of vegetation, especially grasses and

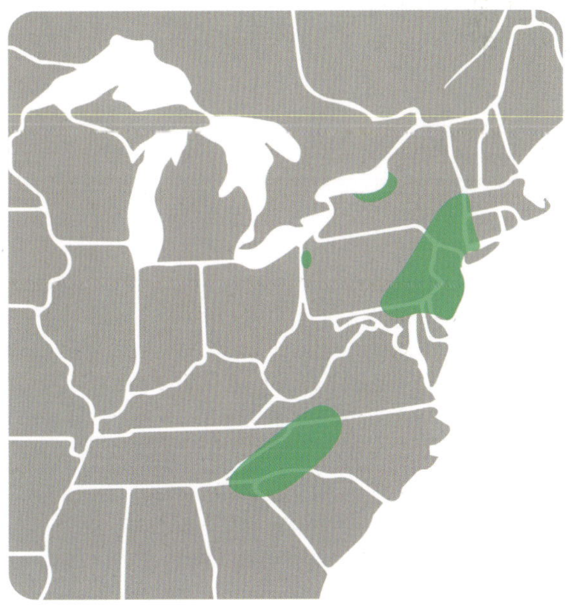

sedges. These include bogs, fens, marshes and wet meadows. Frequently basks but easily overlooked.

DIET Insects, snails, slugs, worms and other invertebrates. Also, occasionally other small animals, carrion and plant material.

REPRODUCTION Lays up to six eggs in a nest dug in damp soil or sphagnum, often in a grassy tussock. Only one clutch per year.

CONSERVATION USA – Threatened, GLOBAL – Critically Endangered. Destruction and pollution of its wetland habitats for agriculture and other development has broken this turtle's large historic range into small, disjunct populations. Illegal collection for the pet trade is also a great risk to this species.

WOOD TURTLE
Glyptemys insculpta

With its bumps, grooves, and ridges, the unique shell of the Wood Turtle almost appears as if it has been carved from wood. This turtle truly walks the line between aquatic and terrestrial and is equally at home on the river bottom or the forest floor.

IDENTIFICATION Carapace is low in profile with a weak central ridge that flattens with age. Texture is very rough, with each scute being convex and covered in grooves and ridges. Older individuals may become somewhat smoother. Overall color is brown or grayish, often with a pattern of radiating yellowish lines. Plastron is yellow with a large, dark patch on each scute. Head and legs are dark, with a varying amount of orange on the undersides of the neck and legs and on the skin in between. Carapace length up to 23 cm (9 in).

SIMILAR SPECIES The unusually textured carapace and orange skin make this turtle nearly unmistakable. Compare to Bog Turtle and Common Box Turtle. Somewhat similar to Diamondback Terrapin but does not share its coastal habitat.

RANGE East coast from Nova Scotia south to Virginia and westward around the Great Lakes to Minnesota and Iowa.

HABITAT Clear rivers and streams with slow currents, sandy bottoms and muddy banks. Often basks, and frequently wanders on land in forested areas, fields and meadows.

DIET Insects, millipedes, slugs, earthworms, other invertebrates, carrion, fungi and plant material. Readily forages both on land and in the water.

REPRODUCTION Lays up to 20 eggs in a nest in soft or sandy soil, sometimes hundreds of meters from the water. Only one clutch per year.

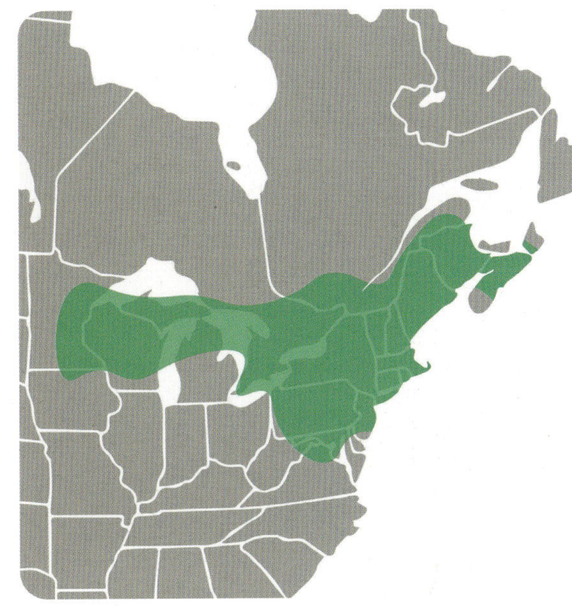

CONSERVATION CANADA – Threatened, USA – Not Listed, GLOBAL – Endangered. Development and alteration of this turtle's riverine and forest habitat may be the primary threat, but it is also threatened by road mortality, illegal collection for the pet trade and nest predation by predators such as raccoons.

A WORM'S WORST NIGHTMARE

Wood Turtles spend much of their time on land and this gives them access to a world of food that their aquatic cousins can only dream of. A breakfast of mushrooms or lunch of millipedes, for example, would be out of reach to a turtle confined to water. There is one terrestrial delight that Wood Turtles especially love, but it can be challenging for even the most high-and-dry turtle to acquire.

Earthworms certainly live in the Wood Turtle's land-based habitats, but they are usually protected by their subterranean habits. Deep in the soil, even the juiciest worm is out of reach to a hungry turtle. The Wood Turtle needs a special trick to coax the worms to the surface, and that trick taps into the worms' biggest fear: moles.

Moles are ferocious predators of worms in their underground homes, and worms are ever wary that death-by-mole may be only seconds away. It turns out that a mole digging towards a worm creates a distinctive vibration. When the worm feels that vibration, its only hope is to get to the surface as fast as possible. This is where the Wood Turtle comes in.

When a Wood Turtle hunts worms, it picks a promising spot on the forest floor, then stomps its feet in a rhythmic, alternating pattern. This pattern mimics the vibration of a digging mole and the worms flee to the surface as quickly as they can. Unfortunately for the worms, what waits aboveground is not a safe, mole-free haven, but the snapping jaws of an excited Wood Turtle.

Photo © Kyle Horner

COMMON BOX TURTLE
Terrapene carolina

With its domed shell, plodding legs and love of terrestrial habitats, the Common Box Turtle is more like a tiny, woodland tortoise than the pond turtles to which it is related. Its incredible, hinged plastron closes completely against the carapace, forming a seal so tight that you could scarcely slide a knife through it.

Photo © Kyle Horner

IDENTIFICATION Carapace is very high-domed and round and may be relatively smooth or have concentric grooves on each scute. Base color is brown or almost black, with a variable, radiating pattern of lines, spots, or blotches, except in Three-toed subspecies (see opposite). Plastron is double-hinged and able to close very tightly against the front and rear edges of the carapace. Plastron color is brown with variable dark markings. Head and legs are brown or grayish and variably marked with yellow or orange. Some males are spectacularly adorned. Carapace length up to 20 cm (8 in).

SIMILAR SPECIES Ornate Box Turtle is the most similar. Also compare to Blanding's Turtle and Wood Turtle.

RANGE In the north, from Massachusetts west across the southern shore of Lake Erie, then up into Michigan. Extends south to the Florida Keys, and west as far as central Kansas, Oklahoma, and Texas.

HABITAT Primarily woodlands, but sometimes swamps, meadows and grasslands. May soak in shallow water but rarely ventures deeper. Often digs into leaf litter or soft soil. May retreat into burrows made by other animals.

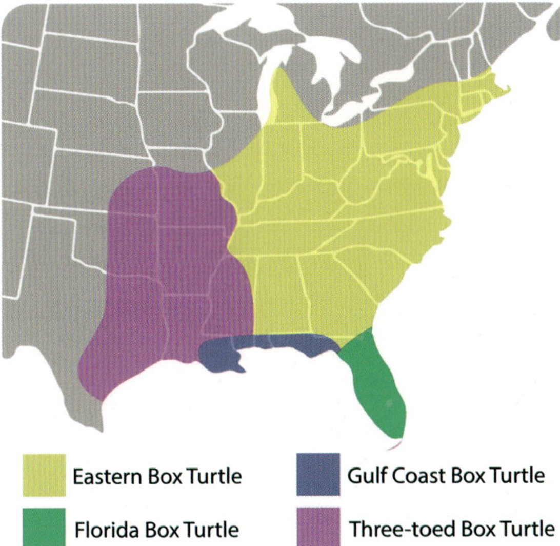

DIET Widely variable, including insects, snails, slugs, earthworms, other invertebrates, carrion, fungi, fruit and flowers.

REPRODUCTION Lays up to 11 eggs in a nest dug in soft or sandy soil. Up to five clutches per year.

SUBSPECIES Divided into six subspecies, four of which occur in Canada and the United States: Eastern Box Turtle (*T. c. carolina*), Florida Box Turtle (*T. c. bauri*), Gulf Coast Box Turtle (*T. c. major*) and Three-toed Box Turtle (*T. c. triunguis*). Debate exists about the taxonomy of this species and there is little consensus. Some sources recognize Three-toed or Florida Box Turtle as separate species and some do not recognize Gulf Coast Box Turtle at all. Subspecies may interbreed where their ranges meet.

CONSERVATION CANADA – Extirpated, USA – Not Listed, GLOBAL – Vulnerable. Archaeological evidence suggests this species may once have lived in extreme southern Ontario, Canada, but the very small number of modern sightings are believed to be of released individuals. Somewhat common but declining across its range. Destruction of forest habitat, road mortality and collection for the pet trade all threaten this turtle.

Eastern Box Turtle has a highly variable appearance, often with a dense pattern of heavy, yellow blotches that may connect into chunky lines.

Photo © Dick Snell

Florida Box Turtle has a pattern of finer lines that radiate from the upper corners of each scute.

Photo © Evan Grimes

Gulf Coast Box Turtle has a variable pattern, rarely as dense as Eastern, but rear margin of carapace is usually strongly flared. Head may have pale blotches.

Three-toed Box Turtle often lacks any pattern, instead being uniform olive or tan. Some dark blotches or faint yellow lines may be visible.

ORNATE BOX TURTLE
Terrapene ornata

Very similar to its more easterly cousin, the Ornate Box Turtle makes its home in the open grasslands of the Great Plains. It once roamed extensively with the American Bison and happily foraged for insects from the poop piles of its prairie companion.

Photo © Evan Grimes

IDENTIFICATION Carapace is very high-domed, often flattened on top, and may be relatively smooth or have concentric grooves on each scute. Base color is dark brown or black, with pattern of yellow lines that radiate from the upper corners of the scutes. Plastron is double-hinged and able to close very tightly against the front and rear edges of the carapace. Plastron pattern is a maze-like network of black and yellow lines. Head and legs are brown or grayish and variably marked with yellow. Carapace length up to 15 cm (6 in).

ALSO KNOWN AS Western Box Turtle

SIMILAR SPECIES Perhaps most similar to the Florida subspecies of Common Box Turtle with which it does not overlap geographically. Compare to other Common Box Turtle subspecies.

RANGE From the southern tip of Lake Michigan south to western Louisiana and west to Wyoming in the north and Arizona in the south.

HABITAT Primarily open grasslands and savannahs, occasionally forest edges. May soak in

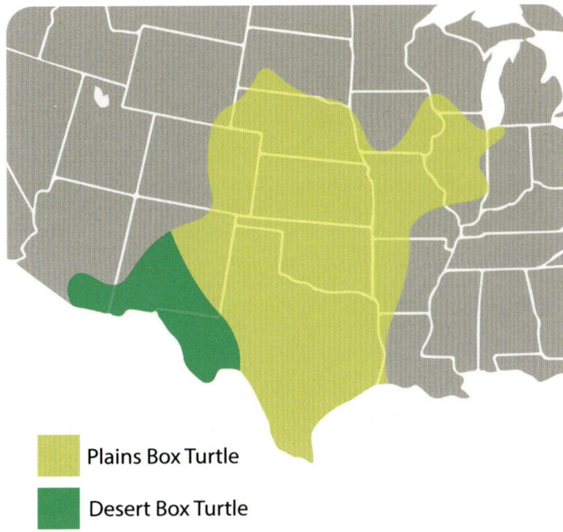

Plains Box Turtle
Photo © Evan Grimes

Desert Box Turtle

Faded Desert Box Turtle

shallow water but rarely ventures deeper. Often shelters under shrubs and may burrow into the soil or retreat into burrows made by other animals.

DIET Widely variable, including insects, slugs, earthworms, other invertebrates, carrion, grasses, cacti, fruit and flowers. Sometimes forages for insects in cow droppings.

REPRODUCTION Lays up to eight eggs in a nest dug in sandy soil. Up to two clutches per year.

SUBSPECIES Divided into two subspecies: Plains Box Turtle (*T. o. ornata*) and Desert Box Turtle (*T. o. luteola*). Desert typically has more and finer lines on the carapace than Plains. Lines on Desert often fade completely to uniform brown with age while Plains usually retains its pattern with age. The subspecies may interbreed where their ranges meet.

CONSERVATION USA – Not Listed, GLOBAL – Near Threatened. Conversion of prairie grassland to agricultural use has significantly impacted this turtle. Road mortality, run-ins with agricultural equipment and collection for the pet trade are also major risk factors.

THE TORTOISES

Most of the world's tortoises are native to tropical and arid environments, and the four species that occur north of Mexico are all found in the very southern parts of the United States. These four closely related turtles are very similar to each other, but there is little overlap in their ranges. Their adaptations to living on land make them visually unlike any other turtle and there should be little confusion when determining whether or not it's a tortoise you're looking at.

Tortoises are built like tanks. All of the species that live in North America have heavy, high-domed carapaces and broad, fixed plastrons. The skin of their heads and legs is thick and scaly. The front legs are particularly well armored. When a tortoise draws its head back into its shell, those robust front legs are pulled in front of the face, creating an impenetrable barrier that protects the tortoise's head.

The protective adaptations of the tortoises likely evolved to keep them safe, not only from predators that would try to eat them, but also from being stepped on by larger land animals. Adult tortoises have few threats to worry about, and this allows them the freedom to roam around at a leisurely pace, munching on leaves, flowers, fruits, cacti and whatever other delicious delights they encounter.

Most of our tortoises are accomplished burrowers and spend much of their time resting deep in their extensive burrows. These subterranean homes protect them from excessive heat, cold, dehydration and even wildfires, which are common in their grassland or scrubland habitats. Mornings and evenings are often the best times to encounter tortoises, when the temperatures are just right for a stroll around the neighborhood.

A TURTLE THAT DOESN'T SWIM

You might think that turtles and water go together, but the tortoise's heavy shell, blocky shape, stumpy legs and un-webbed feet could not be more poorly adapted to aquatic activities. The awkward truth is that a tortoise is a turtle that can't swim a stroke.

A tortoise may wade into shallow water for a drink or a soak, but if you placed one in the deep end of a pond or river it would simply walk along the bottom to the shore. If it was unable to amble to the shallows, it would be in grave danger of drowning.

Evolution is all about compromise and, very often, being good at one thing means being bad at another. You'll never see a tortoise doing the dog-paddle across a pond but they are perfectly built for plodding through soft sand, bulldozing into dense thickets or digging deep below the desert floor. They have evolved to leave the water behind and there's no looking back.

OPPOSITE A Gopher Tortoise in safe mode.

GOPHER TORTOISE
Gopherus polyphemus

An emblematic resident of pine-dominated flatwoods and scrubland in the American southeast, the Gopher Tortoise is the only tortoise species found east of Texas. Its burrows are a prominent feature of these habitats and provide a home for many other species.

IDENTIFICATION Carapace is high-domed, robust, elongate and sometimes flattened on top. Color is brown, tan or gray. The scutes of young tortoises may have a pale central spot. Plastron is large, solid and yellowish in color with or without darker smudging. Head and legs are roughly the same color as the carapace. Legs are elephantine and feet are not webbed. Carapace length up to 38 cm (15 in).

SIMILAR SPECIES Other North American tortoises are very similar but don't overlap its range. Compare to Common Box Turtle.

RANGE From southern South Carolina to nearly the tip of Florida and west as far as eastern Louisiana.

HABITAT Open uplands with sandy soils, including flatwoods, savannahs, sandhills, prairie,

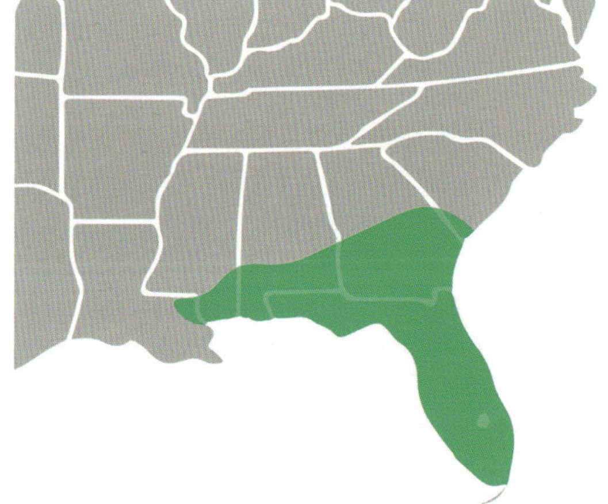

grasslands and dunes. Often associated with pine trees. Constructs deep burrows in which it frequently shelters.

DIET Primarily plant material, including grasses, leaves, needles, cacti, shoots, fruit, seeds and flowers. Occasionally eats fungi, carrion, insects and even the droppings of other animals.

REPRODUCTION Lays up to 13 eggs in a nest typically dug in the soft sand around the entrance of its burrow. Only one clutch per year.

CONSERVATION USA – Threatened, GLOBAL – Vulnerable. Loss and alteration of its

sandy habitat for development, agriculture and other uses may be the biggest risk to this species. Road mortality and illegal harvesting for food or the pet trade are also significant threats.

A BURROW FOR ONE, A BURROW FOR ALL

A Gopher Tortoise burrow is not just a little hole in the ground. It may be over 10 m (30 ft) long and reach depths of up to 3 m (10 ft). A tortoise may dig multiple burrows throughout its life, moving periodically from one to another. These subterranean refuges provide the tortoise with an environment that stays cool when the weather is hot, warm when the weather is cold and humid when the weather is dry.

Good places to live are rare in challenging environments and Gopher Tortoises are rarely the only residents in their burrows. Over 350 species use the tortoises' homes, including rodents, rabbits, snakes, lizards, frogs and many others. These creatures use the burrows for protection from predators, the elements and wildfires, and they may occupy a burrow at the same time as the tortoise, or after it has been abandoned by the tortoise.

By creating burrows, a Gopher Tortoise becomes an important architect of its environment. The animals that live alongside it have evolved to depend on what the tortoise provides. Several rare species — like the threatened Eastern Indigo Snake and the aptly-named Gopher Frog — rely on the burrows for their survival. This interdependency makes the Gopher Tortoise a keystone species, and one whose conservation benefits hundreds of others.

Gopher Tortoise and burrow

Indigo Snake

Gopher Frog

TEXAS TORTOISE
Gopherus berlandieri

The smallest and roundest of the North American tortoises, the Texas Tortoise is found only in southern Texas and northern Mexico. Unlike our other tortoises, this species rarely burrows, preferring to simply hide under vegetation or wedge itself between rocks for protection.

IDENTIFICATION Carapace is high-domed, robust, and almost as wide as it is long. Color is brown or gray. Scutes of young tortoises may have a pale central spot. Plastron is large, solid, yellowish in color and often heavily marked with black. Head and legs are roughly the same color as the carapace. Legs are elephantine and feet are not webbed. Carapace length up to 38 cm (15 in).

ALSO KNOWN AS Berlandier's Tortoise

SIMILAR SPECIES Other North American tortoises are very similar but don't overlap its range. Compare to Common and Ornate Box Turtles.

RANGE Endemic to southern Texas.

HABITAT Dry, sandy, open habitats including grassland and scrub. Does not often burrow, instead simply pushing its way under a shrub, grass tussock or cactus for shelter and protection.

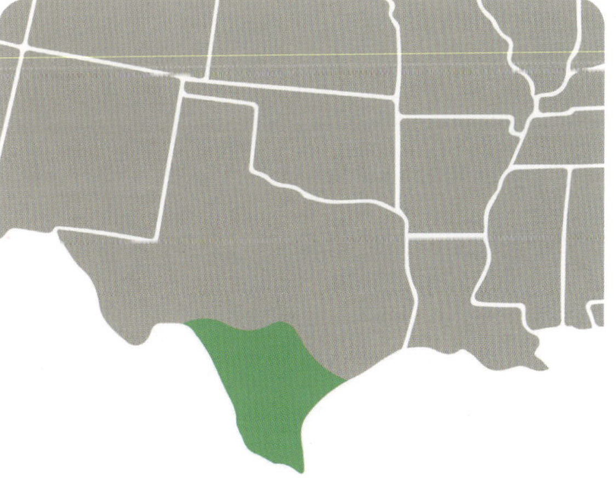

DIET Primarily plants, especially cactus pads and fruit. Occasionally eats carrion, insects, and even the droppings of other animals.

REPRODUCTION Lays up to five eggs in a nest or split between multiple nests dug in sandy soil. Only one clutch per year.

CONSERVATION USA – Not Listed, GLOBAL – Least Concern. While not federally or globally listed as at-risk, both the state of Texas and country of Mexico recognize this species as Threatened. Loss and alteration of its habitat, road mortality and illegal harvesting are all threats to this tortoise.

SONORAN DESERT TORTOISE
Gopherus morafkai

Until recently, the Sonoran Desert Tortoise was considered the same species as its close cousin the Mojave Desert Tortoise. The two are extremely similar, but if you're standing east of the Colorado River, you're in Sonoran territory.

IDENTIFICATION Carapace is moderately to highly domed, robust and somewhat pear-shaped, being narrower at the front and wider at the rear. Color is brown, gray, or tan. The scutes of young tortoises may have a pale central spot. Plastron is large and solid, yellowish in color and may be with or without darker smudging. Head and legs are roughly the same color as the carapace. Legs are elephantine and feet are not webbed. Carapace length up to 38 cm (15 in).

ALSO KNOWN AS Morafka's Desert Tortoise

SIMILAR SPECIES Very similar to Mojave Desert Tortoise. Perhaps best identified by range, but it may be possible to differentiate based on the shape of the carapace.

RANGE The Sonoran Desert in Arizona, east of the Colorado River.

HABITAT Rocky desert hillsides and slopes covered in dry grassland or scrubby vegetation. Constructs deep burrows in compact, sandy soil.

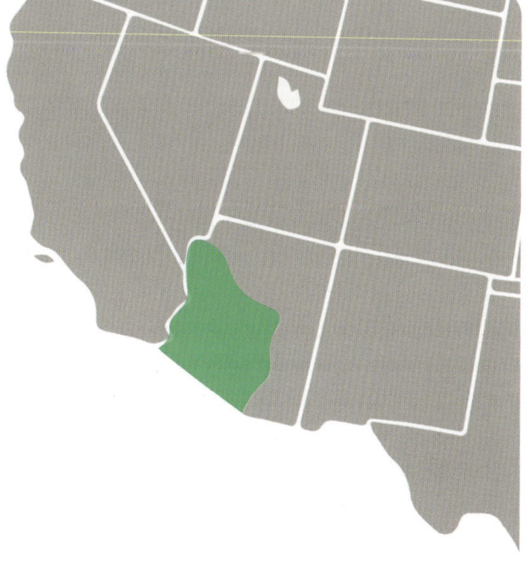

Photo © Grover Brown

DIET Primarily plants, including leaves, grasses, cacti, fruit and flowers. Occasionally eats carrion, insects and even the droppings of other animals.

REPRODUCTION Lays up to 12 eggs per year in a nest dug in sandy soil, usually near the burrow entrance. Only one clutch per year.

CONSERVATION USA – Not Listed, GLOBAL – Not Assessed. Because this species was only recently separated from its close cousin the Mojave Desert Tortoise, it has not yet been separately listed as at-risk either in the United States or globally. Future assessments will likely do so, as this tortoise is threatened by habitat destruction, road mortality and illegal collection.

MOJAVE DESERT TORTOISE
Gopherus agassizii

Nearly identical to its Sonoran counterpart, the Mojave Desert Tortoise resides north and west of the Colorado River. Whereas its close cousin likes desert slopes and hillsides, you're more likely to encounter a Mojave in the valleys and washes below high ground.

IDENTIFICATION Carapace is moderately to highly domed, robust and uniformly wide. Color is brown, gray or tan. The scutes of young tortoises may have a pale central spot. Plastron is large, solid and yellowish in color with or without darker smudging. Head and legs are roughly the same color as the carapace. Legs are elephantine and feet are not webbed. Carapace length up to 22 cm (8.5 in).

ALSO KNOWN AS Agassiz's Desert Tortoise

SIMILAR SPECIES Very similar to Sonoran Desert Tortoise. Perhaps best identified by range, but it may be possible to differentiate based on the shape of the carapace.

Photo © Dawn Nelson

RANGE The Sonoran and Mojave deserts west and north of the Colorado River in California, Nevada, northwestern Arizona and southwestern Utah.

HABITAT Desert valleys, lowlands, and washes with scrubby vegetation, yucca, or cacti. Constructs deep burrows in compact, sandy soil.

DIET Primarily plants, including leaves, grasses, cacti, fruit and flowers. Occasionally eats carrion, insects and even the droppings of other animals.

REPRODUCTION Lays up to 16 eggs per year in nest dug in sandy soil, usually near the burrow entrance. Up to three clutches per year.

CONSERVATION USA – Threatened, GLOBAL – Critically Endangered. Habitat destruction due to urban development, agriculture, and other land uses is likely the greatest threat to this tortoise. Illegal harvesting, mortality on roads, and collisions with offroad vehicles are also substantial risks.

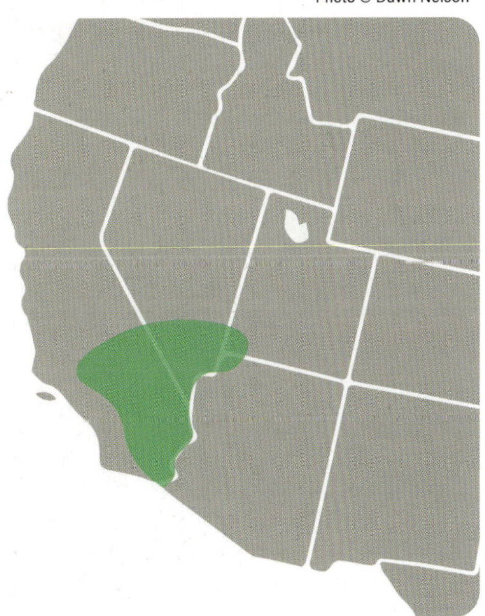

Family Trionychidae
THE SOFTSHELLS
Spiny Softshell
Florida Softshell
Smooth Softshell

Photo © Carl Franklin – TexasTurtles.org

t's hard to picture a more unusual turtle than a softshell. Their namesake feature makes them the antithesis of what we imagine a turtle to be, and their strangeness doesn't stop there. There are just three species in North America and they are all very similar in appearance.

The soft shell is surely the first thing you'll notice on a softshell and it is just as soft and pliable as it looks. The leathery skin covers a bony structure but the bone does not extend to the margin, so the edges are flappy and flexible. There are no hard scutes providing protection, just a supple surface of smooth skin. The shell is very low and flat in profile and comparisons to a pancake are all but inevitable.

OPPOSITE Flexibility is the name of the game for softshell turtles.

The plastron is as soft as the carapace and doesn't extend far to the rear, so the hind legs are fully exposed at all times. The legs are widely splayed and the feet are very large and strongly webbed. These powerful paddles combine with the turtle's sleek, stream-lined shape to make the softshell a fast, dynamic swimmer.

A softshell's face may be the strangest feature of all, and it looks positively alien at first glance. The smooth skin wraps tightly around beady eyes and a powerful beak while the face terminates in a long, tapering, pig-like nose. This built-in snorkel allows the softshell to bury itself in the sand or mud at the bottom of a shallow river or wetland, then stretch its long neck upwards to breathe with only the tip of its nose breaking the water's surface.

The softshells' unique adaptations make them fast in the water, but these turtles acquit themselves well on land, too. Their light, flexible armor allows them to do something few turtles can do: run. A startled softshell can achieve surprising speed over short distances on land, reportedly even outrunning researchers who are trying to capture them.

This speed serves the softshells well, as they love to bask on exposed sandbars and shorelines where they would be an easy meal for a predator if caught. A softshell may look relaxed as it soaks in the sun, but its flight mode is on a hair trigger. The slightest suggestion of danger will send it dashing for the water. Most turtles sacrifice speed for safety but, for the softshells, speed and safety are inexorably linked.

Unusual in every way, the softshells seem to break all the rules of what it means to be a turtle.

SPINY SOFTSHELL
Apalone spinifera

The name of the Spiny Softshell may conjure an impressively armored turtle, but the so-called spines are simply small, conical bumps on the front edge of the carapace. This identifying feature can be difficult to see in the field, especially on older females.

Photo © Eric Gofreed

IDENTIFICATION Carapace is leathery, flexible and very flat in profile. Front edge has many small, conical projections. Color is brown, olive or tan, with black or white spots or rings. Pattern is often obscured in older females, which appear uniform or mottled. Plastron is also soft and does not extend far towards the rear. Plastron color is white or yellowish. Head and legs are roughly the same color as the carapace and may be covered in black spots. Pale lines with black borders run along the face and neck. Carapace length up to 52 cm (20 in).

SIMILAR SPECIES Compare to Smooth and Florida Softshells.

RANGE Widespread and somewhat scattered depending on availability of riverine habitat. From the eastern edge of the Great Lakes south to North Carolina, westward to Wyoming, Colorado and New Mexico, continuing west along the southern border to California and Nevada. Disjunct populations occur in Quebec, several northeastern states and Montana. Introduced populations exist in California and Washington.

HABITAT Primarily rivers and large streams, but sometimes lakes and other bodies of water. Prefers shallow areas with soft mud or sandy bottoms where it may bury itself. Regularly basks, either by hauling out of the water or simply floating at the surface. Rarely found far from water.

Eastern Spiny Softshell has carapace with pale rim and black border. Black spots cover the carapace, often forming hollow, eye-like spots towards the center.

Gulf Coast Spiny Softshell has carapace with pale rim and at least two dark, parallel borders along the rear margin. Black spots cover the carapace and may become hollow towards the center.

Pallid Spiny Softshell has carapace with white spots not ringed with black. White spots only or primarily on the back half of the carapace.

Guadalupe Spiny Softshell has carapace with indistinct border and is fully covered in white spots surrounded by black rings.

Texas Spiny Softshell has carapace with pale border which is notably widest along rear margin. White spots present only on rear third of the carapace.

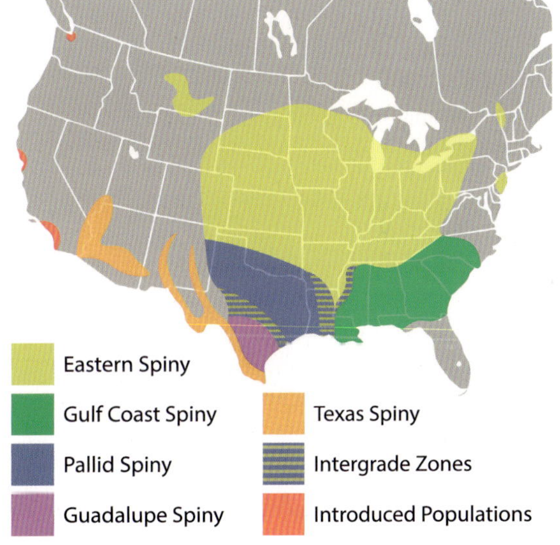

DIET Insects, crayfish, other invertebrates, fish and carrion.

REPRODUCTION Lays up to 35 eggs in a nest dug in sand or gravel not far from water, typically in a riverbank. Up to two clutches per year.

SUBSPECIES Divided into five subspecies: Eastern (*A. s. spinifera*), Gulf Coast (*A. s. aspera*), Pallid (*A. s. pallida*), Guadalupe (*A. s. guadalupensis*), and Texas (*A. s. emoryi*). Subspecies may interbreed in several substantial intergrade zones.

CONSERVATION CANADA – Endangered, USA – Not Listed, GLOBAL – Least Concern. This species has a small, restricted range in Canada, where alteration and disturbance of its limited habitat is a substantial risk. Elsewhere it is still relatively common but is threatened by harvesting for food (both for local and overseas markets), habitat destruction and pollution.

FLORIDA SOFTSHELL
Apalone ferox

The largest and darkest of our softshells, the Florida Softshell also differs from its cousins in its choice of habitat. Eschewing the rivers beloved of its spiny and smooth family members, this species can be found in almost any non-moving body of fresh water.

IDENTIFICATION Carapace is leathery, flexible and very flat in profile. Front edge has many small, round bumps. Color is dark brown or dark greenish with little or no pattern. Plastron is also soft and does not extend far towards the rear. Plastron color is white or yellowish. Head and legs are roughly the same color as the carapace and usually unmarked, although young individuals may have some pale markings. Carapace length up to 63 cm (25 in).

SIMILAR SPECIES Compare to Spiny and Smooth Softshell.

RANGE Southern Alabama, Georgia, and South Carolina, south to the Florida Keys.

HABITAT Primarily bodies of still water, including lakes, ponds, canals, marshes and ditches. Occasionally slow-moving rivers.

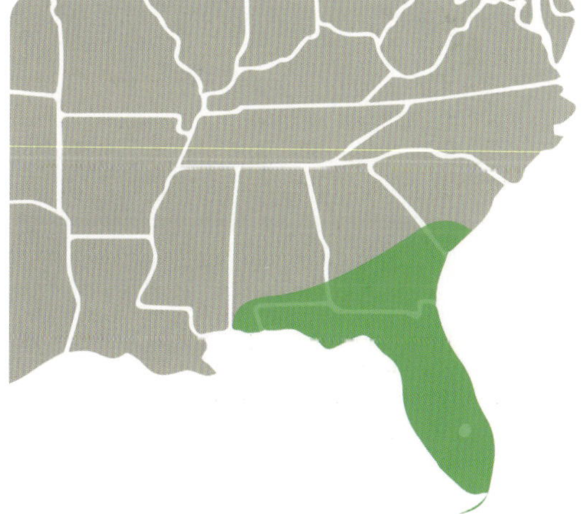

DIET Snails, insects, other invertebrates, fish, amphibians, carrion and occasionally snakes and small birds. May occasionally consume plant material.

REPRODUCTION Lays up to 38 eggs in a nest dug in sandy soil, usually not far from the water. Up to six clutches per year.

CONSERVATION USA – Not Listed, GLOBAL – Least Concern. Relatively common throughout its range but threatened by habitat loss and harvesting for food.

SMOOTH SOFTSHELL
Apalone mutica

The aptly named Smooth Softshell is the smallest and flattest of the North American softshells. It lacks the projections on the carapace that can be found in both of our other softshell species. This turtle is almost exclusively found in rivers and particularly loves clean, fast-flowing habitats.

Photo © Grover Brown

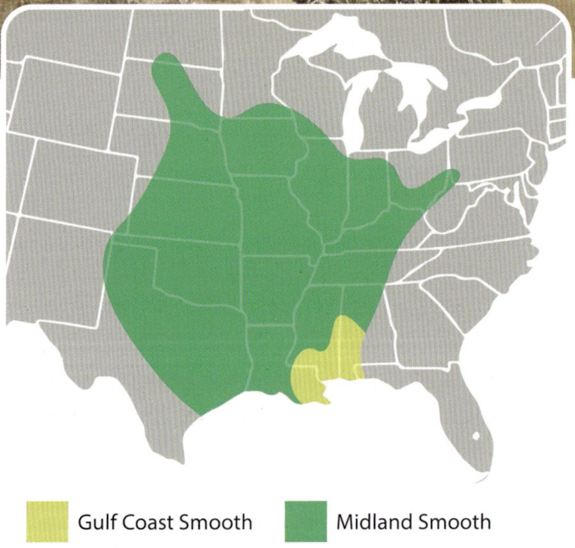

IDENTIFICATION Carapace is leathery, flexible and very flat in profile. Front edge has no bumps or projections. Color is brown, olive or grayish with a pale border and scattered, dark spots. Pattern is often obscured in older females, which appear uniform or mottled. Plastron is also soft and does not extend far towards the rear. Plastron color is white or yellowish. Head and legs are roughly the same color as the carapace and usually unmarked, aside from pale lines with dark borders that run along the face and neck. Carapace length up to 35 cm (14 in).

SIMILAR SPECIES Compare to Spiny and Florida Softshells.

RANGE Widespread and somewhat scattered depending on availability of riverine habitat. From southwestern Pennsylvania south to the Florida panhandle and west as far as North Dakota in the north and eastern New Mexico in the south.

Photo © Peter Paplanus

Gulf Coast Smooth Softshell

Photo © Grover Brown

Midland Smooth Softshell

Photo © Dr. Joe Coelho

HABITAT Primarily large, clear rivers and streams with fast currents, only occasionally lakes and other bodies of water. Prefers shallow areas with soft mud or sandy bottoms where it may bury itself. Regularly basks, either by hauling out of the water or simply floating at the surface. Rarely found far from water.

DIET Insects, crayfish, snails, other invertebrates, fish, amphibians and carrion. Occasionally eats some plant material.

REPRODUCTION Lays up to 33 eggs in a nest dug in sandy soil, usually quite close to the water. Only lays one clutch per year.

SUBSPECIES Divided into two subspecies: Midland Smooth Softshell (*A. m. mutica*) and Gulf Coast Smooth Softshell (*A. m. calvata*). Midland typically has a white line passing through the eye, extending forward towards the snout, although often faintly. White line behind eye has narrow black borders. Gulf Coast does not have white line between the eye and snout. White line behind eye has thick black borders. Subspecies may interbreed where ranges meet.

CONSERVATION USA - Not Listed, GLOBAL - Least Concern. Relatively common throughout its range, but threatened by habitat loss, pollution and harvesting for food.

Families Dermochelyidae and Cheloniidae

THE SEA TURTLES

Leatherback Sea Turtle
Green Sea Turtle
Loggerhead Sea Turtle
Hawksbill Sea Turtle
Kemp's Ridley Sea Turtle
Olive Ridley Sea Turtle

Sea turtles are well known and well loved all over the world, even in places far from their oceanic habitats. This famous group of turtles actually consists of two families, with most belonging to the Cheloniidae and the unique Leatherback Sea Turtle being the only member of its own family, Dermochelyidae.

The sea turtles live nearly their entire lives in the ocean and their bodies are designed for this pelagic lifestyle. Sea turtle shells are relatively flat in profile and pointed at the end, making them perfectly hydrodynamic. These turtles are unable to pull their heads, legs or tails inside their shells for protection. The Leatherback Sea Turtle scarcely has a shell at all, lacking both hard scutes on the outside and a solid, bony structure on the inside. This flexible armor adapts to varying ocean pressures, allowing the Leatherback to dive deeper than any other turtle.

The legs and feet of sea turtles may be their most distinctive and unusual feature, as they are completely modified into seal-like flippers. These dedicated limbs make sea turtles unrivalled swimmers among their kin, but they are nearly useless on land. Fortunately, they are rarely required to walk, as these turtles leave the water only briefly to deposit their eggs on their nesting beaches.

Sea turtles are mostly tropical in their distribution, but some species migrate to temperate waters during warm seasons to take advantage of productive feeding grounds. Of the six species that can be seen in North American waters, several venture as far north as the Canadian coast. Nesting occurs exclusively on sandy beaches in hot places, and some species nest on the southernmost beaches of the United States.

A DAY AT THE BEACH, A LIFETIME AT SEA

Sea turtles are dedicated beach bums, and you'll almost certainly never see one in any other terrestrial environment. Soft, sandy beaches provide the perfect substrate for digging nests, and their proximity to water allows these specialized turtles to spend as little time as possible away from their marine habitat.

Sea turtles are picky about their beaches, and some migrate great distances to nest on the perfect one. Many females will even return to the specific beaches they hatched on. Males and females meet and mate in the ocean, usually near the nesting grounds, then the females drag themselves onto the beaches to dig their nests and lay their eggs.

A female sea turtle may make multiple nests in a season, after which she leaves the beach and may not return for several years. Like most turtles, she leaves her young to fend for themselves. Baby sea turtles may have the most perilous first day of all turtles, as they must emerge from their nests and dash for the water across an open beach full of eager predators. Crabs, gulls, crows, raccoons, coyotes and many other animals have evolved to take advantage of this seasonal feast.

If they reach the water successfully, conditions become only slightly safer. Little is known about the lives of juvenile sea turtles, as they are rarely seen after leaving the beach, but it seems that many head straight for deep water and take up residence in floating mats of seaweed which provide both food and shelter. Male sea turtles will never leave the ocean again, while females will spend a decade or even two at sea before reaching maturity and returning to their nesting beaches to lay eggs of their own.

ABOVE A day at the beach is a rare occurrence for a sea turtle, but one of great importance.

LEATHERBACK SEA TURTLE
Dermochelys coriacea

The Leatherback Sea Turtle is an extreme specialist in every way, from its deep ocean diving to its diet of almost exclusively jellyfish. It is also the largest turtle in the world, growing to be nearly as long and as heavy as a small car.

IDENTIFICATION Carapace is leathery, flat in profile and with no scutes. Front margin is rounded, and rear margin comes to a point, creating a teardrop shape. Seven ridges run the length of the carapace. Plastron is also leathery and somewhat indistinct. Color of the turtle's entire upper surface is dark gray or black with scattered white spots. Entire underside is white with variable dark blotches. Exceptionally large individuals may reach a carapace length over 2 m (78 in) and weight over 600 kg (1,322 lbs), though reports of the biggest Leatherbacks vary in their credibility.

ALSO KNOWN AS Lute Turtle or Luth.

SIMILAR SPECIES The unique shell of this species makes it unmistakable.

RANGE Worldwide. May occur in nearly all ocean waters adjacent to Canada and the United States, as far north as Newfoundland in the Atlantic and Alaska in the Pacific. Some nesting occurs on Florida beaches.

Photo © Cheng Sung Hua

HABITAT Almost exclusively the open ocean as adults. Juvenile Leatherbacks are rarely encountered, but may occasionally inhabit shallower, warmer waters.

DIET Almost entirely jellyfish, but very occasionally other soft-bodied invertebrates such as tunicates and squid.

REPRODUCTION Lays up to 110 eggs in a nest dug in beach sand. May lay up to 10 clutches in a nesting season, but typically nests only every two or three years.

CONSERVATION CANADA – Endangered, USA – Endangered, GLOBAL – Vulnerable. In the open ocean, these turtles are at risk of accidental capture by fishing operations, vessel strikes and the ingestion of marine plastics, especially plastic bags which resemble the jellyfish they eat. On the nesting beaches they are threatened by development and harvesting of both eggs and adult turtles for consumption.

A VERY PICKY EATER

The Leatherback Sea Turtle is undoubtedly the most selective eater of all North American turtles, as the vast majority of its diet comprises just a single thing: jellyfish. These soft, squishy bags of water are an unlikely dietary preference for such an enormous animal, and an adult Leatherback may need to consume hundreds of kilograms of jellyfish per day just to maintain its massive bulk.

This specialization on jellyfish as food is the reason for most everything that makes the Leatherback unique. The turtle's bizarre shell is not underpinned by a solid layer of bone, but by a network of small, bony plates called osteoderms. This structural layer is flexible, allowing the Leatherback's shell to expand and contract in response to varying ocean pressure. So equipped, the Leatherback can plunge to depths of over 1,200 m (3,937 ft) in search of its gelatinous dinner.

These occasional deep dives likely help the Leatherback locate its prey, which retreats to deep waters during the day, but most feeding occurs

Jellyfish

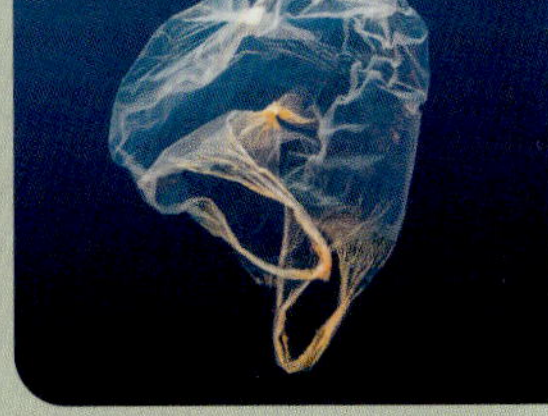
Plastic bag

at night when the jellyfish rise to the surface. By waiting until after dusk, the turtle can consume large quantities of jellyfish without the energy cost of making repeated deep dives. Leatherbacks see well in the dark, allowing them to both scout the depths and feast at the surface under the stars.

This unusual feeding strategy makes the Leatherback uniquely susceptible to a particular danger: ocean plastics. A plastic grocery bag adrift in the water after dark looks remarkably like a jellyfish to a hungry turtle, and ingestion of plastic can be fatal. Reduction and careful disposal of our plastic waste may be critical to the survival of the leviathan Leatherback.

GREEN SEA TURTLE
Chelonia mydas

The misleading name of the Green Sea Turtle apparently refers to the color of the fat found under its shell. The carapace is much more intricate, colorful and beautiful than the name suggests. Perhaps the word green in its name should actually refer to the turtle's diet, which is uniquely herbivorous for a sea turtle.

IDENTIFICATION Carapace is flat in profile and roughly oval in shape. Five large scutes run down the centre, with four more on either side. Color is a variable combination of green, brown and yellow, often with a radiating pattern on each scute. Plastron is yellowish. Head is relatively small with a short, blunt snout. Each front flipper has a single claw. Carapace length up to 150 cm (59 in). Weight may exceed 300 kg (660 lbs).

ALSO KNOWN AS Green Turtle

SIMILAR SPECIES Compare to all other sea turtles except the distinctive Leatherback.

RANGE Worldwide. In the Atlantic, as far north as the Canadian Maritimes. In the Pacific, may reach Alaskan waters. Nests in Florida and sometimes as far north as the Carolinas.

HABITAT Typically shallow ocean habitats near coastlines, including bays, lagoons, reefs and salt-marshes. Especially fond of seagrass beds. Juvenile turtles may use deeper ocean waters.

DIET As adults, mostly seagrasses and algae. Young turtles eat a variety of marine invertebrates including crustaceans, worms and jellyfish.

REPRODUCTION Lays up to 125 eggs in a nest dug in beach sand. May lay up to six clutches in a nesting season, but typically nests only every two to four years.

CONSERVATION **USA** – Endangered, **GLOBAL** – **Endangered.** Harvest of both eggs and adults from nesting beaches for food is a major threat to this species. Accidental catch by fishing operations, alteration of their nesting beaches and degradation of their coastal feeding grounds are also significant risk factors.

FINDING A WAY HOME

Sea turtles perform some of the most incredible migrations on Earth. The highly nomadic Leatherback Sea Turtle may journey over 6,000 km (3,700 mi) between its feeding grounds off the coast of Newfoundland and its nesting beaches in South America. The distances covered by these oceanic wanderers are immense, but it is their navigational abilities that have most impressed researchers.

One of the most significant nesting beaches for Green Sea Turtles in the Atlantic is on Ascension Island. Thousands of turtles migrate every year from waters off the coast of Brazil to this tiny dot in the middle of the South Atlantic, traversing over 2,000 km (1,250 mi) of featureless ocean to land on an island that is just 10 km (6 mi) wide. It is a feat roughly akin to standing on the 50-yard line of a football field and trying to land a nickel on a saucer somewhere in the endzone, blindfolded.

Exactly how sea turtles perform these navigational miracles has long baffled scientists, and extensive research has not yet provided a complete answer. It seems likely that the turtles are able to detect changes in the Earth's magnetic field and use these slight variations to orient themselves. There is some evidence that they also use smell as they get close to their destination. Whatever the mechanism, the Green Sea Turtle clearly has an internal compass that rivals the most sophisticated GPS.

LOGGERHEAD SEA TURTLE
Caretta caretta

Named for its big, blocky head with powerful jaw muscles, the Loggerhead Sea Turtle is well equipped to crack into even the most hard-shelled ocean creatures. This gluttonous sea turtle is no picky eater, and it boasts a larger menu than any other species.

IDENTIFICATION Carapace is flat in profile and roughly heart-shaped. Five large scutes run down the centre, with five more on either side. Color is reddish brown to orange, sometimes with a radiating pattern on each scute. Plastron is white, yellowish or grayish. Head is large and blocky. Each front flipper has two claws. Carapace length up to 120 cm (47 in). Weight may exceed 200 kg (440 lbs).

SIMILAR SPECIES Compare to all other sea turtles except the distinctive Leatherback.

RANGE Worldwide. In the Atlantic, as far north as the Canadian Maritimes and possibly Newfoundland. In the Pacific, may just reach Canadian waters. Nests in Florida and Georgia.

HABITAT Often bays, lagoons, estuaries, and other shallow habitats near shore, but will also venture to deeper waters.

DIET Widely varied. Mostly invertebrates that live on the sea floor, including crabs, horseshoe crabs,

clams, mussels, conches, barnacles, corals, worms, starfish, urchins, anemones, sea cucumbers, squid and many more.

REPRODUCTION Lays up to 125 eggs in a nest dug in beach sand. May lay up to six clutches in a nesting season, but typically nests only every two to four years.

CONSERVATION **CANADA** – Endangered, **USA** – Endangered, **GLOBAL** – Vulnerable. Harvest of both eggs and adults from nesting beaches for food is a major threat to this species. Accidental catch by fishing operations, ingestion of marine plastics, alteration of their nesting beaches and degradation of their coastal feeding grounds are also significant risk factors.

HAWKSBILL SEA TURTLE
Eretmochelys imbricata

Distinguished by its slightly hooked beak and the serrated rear margin of its carapace, the Hawksbill Sea Turtle has unusual taste in food. The sea sponges it mostly eats are both toxic and full of spines, but like a person devouring a plate of hot wings, the Hawksbill is unbothered by its dinner's defenses.

IDENTIFICATION Carapace is flat in profile and roughly heart-shaped. Scutes of the rear margin overlap each other in a serrated pattern. Five large scutes run down the center, with four more on either side. Color is orange, yellow, or brown with a pattern of light and dark streaks. Plastron is yellowish. Head is long and tapering, beak is slightly hooked downward. Each front flipper has two claws. Carapace length up to 95 cm (37 in). Weight may exceed 100 kg (220 lbs).

SIMILAR SPECIES Compare to all other sea turtles except the distinctive Leatherback.

RANGE Mostly tropical waters worldwide. In the Atlantic, ranges as far north as the New England states. In the Pacific, just reaches the waters of California. Nests in Florida.

HABITAT Most often found around coral reefs and in other shallow, coastal habitats such as

lagoons and estuaries. Juvenile turtles are found in deeper waters.

DIET Often primarily sponges, but also anemones, jellyfish, crustaceans, and other invertebrates.

REPRODUCTION Lays up to 200 eggs in a nest dug in beach sand. May lay up to six clutches in a nesting season, but typically nests only every two to four years.

CONSERVATION USA – Endangered, GLOBAL – Critically Endangered. Historical harvesting for the manufacture of tortoiseshell, now banned, drastically impacted this species. Today, harvest of eggs and adults for food, accidental catch by fishing operations, alteration of their nesting beaches and destruction of their coral reef habitats are also significant risk factors.

KEMP'S RIDLEY SEA TURTLE
Lepidochelys kempii

The Kemp's Ridley Sea Turtle has the smallest range of any sea turtle in the world, with adults rarely leaving the Gulf of Mexico. Younger turtles sometimes venture up the east coast of the United States and Canada and have been seen as far north as Newfoundland.

Photo © Heather Cooley

IDENTIFICATION Carapace is flat in profile, rounded, almost as wide as it is long, and roughly heart-shaped. Five large scutes run down the center, with five more on either side. Color is uniformly grayish green. Plastron is whitish or yellowish. Head is relatively small and triangular when seen from above. Each front flipper has one claw. Carapace length up to 75 cm (30 in). Weight may exceed 50 kg (110 lbs).

ALSO KNOWN AS Atlantic Ridley Sea Turtle

SIMILAR SPECIES Compare to all other sea turtles except the distinctive Leatherback.

RANGE Found only in the Atlantic Ocean. Adults live mostly in the Gulf of Mexico, but younger turtles range as far north as the New England states, and rarely even the Canadian Maritimes or Newfoundland. Most nesting occurs on a single beach in Mexico, but some nest on the American Gulf Coast.

HABITAT Mostly shallow waters of the Gulf of Mexico.

DIET Mostly hard-shelled prey including crabs, mussels, urchins, clams and shrimp, but also jelly-fish, fish, squid and algae.

REPRODUCTION Lays up to 110 eggs in a nest dug in beach sand. Many females nest together in groups called *arribadas*. May lay up to four clutches in a nesting season and may nest every one to three years.

CONSERVATION USA – Endangered, GLOBAL – Critically Endangered. Accidental catch by fishing operations in the Gulf of Mexico is perhaps the most significant threat to this turtle. Harvest of eggs and adults for food, marine pollution, oil spills, alteration of their nesting beaches and deg-radation of their coastal feeding grounds are also significant risk factors.

Unlike its rarely roaming cousin, the Olive Ridley Sea Turtle can be found in nearly all tropical areas of the world in the Pacific, Indian, and Atlantic Oceans. In the United States and Canada, they are only seen in the warmer, southern waters of the Pacific Coast.

IDENTIFICATION Carapace is somewhat deeper in profile than other sea turtles, broad, rounded and roughly heart-shaped. At least five large scutes run down the center, with at least six more on either side. Color is uniformly olive green. Plastron is whitish or yellowish. Head is relatively small and triangular when seen from above. Each front flipper has one or two claws. Carapace length up to 75 cm (30 in). Weight may exceed 50 kg (110 lbs).

ALSO KNOWN AS Pacific Ridley Sea Turtle

SIMILAR SPECIES Compare to all other sea turtles except the distinctive Leatherback.

RANGE Tropical waters worldwide. Does not reach the United States in the Atlantic. In the Pacific, ranges as far north as southern British Columbia.

HABITAT Often seen in shallow waters close to shore, but also traverses great distances in the open ocean.

DIET Many marine invertebrates, including crabs, shrimp, jellyfish, tunicates, urchins, mollusks and others. Occasionally fish and fish eggs.

REPRODUCTION Lays up to 110 eggs in a nest dug in beach sand. Many females nest together in groups called *arribadas*. May lay up to three clutches in a nesting season, and often nests every year.

CONSERVATION USA – Endangered, GLOBAL – Vulnerable. Harvest of both eggs and adults from nesting beaches for food is perhaps the biggest threat to this species. Accidental catch by fishing operations and alteration of their nesting beaches are also significant risk factors.

GLOSSARY

AESTIVATION A period of dormancy, somewhat like hibernation, that occurs during the summer. Some turtles in hot or dry environments aestivate by burying themselves in mud or soil, slowing their body processes and waiting for rain or cooler temperatures.

AQUATIC Living primarily in water. Usually refers to freshwater, in contrast to marine.

BASKING The act of lying in the sun, usually to get warm or dry. Many turtles bask by hauling out of the water onto a log or vegetation, while some simply float at the surface.

BRACKISH Slightly salty or having salinity between fresh and saltwater. Brackish water typically occurs where fresh and saltwater meet, as they do at river mouths and in coastal lagoons.

CARAPACE The upper portion of a turtle's shell.

CARNIVOROUS Eating only meat.

CARRION Dead animals, particularly when they are eaten by other animals.

CLOACA The single opening on the underside of a turtle's tail, used for both pooping and mating.

CLOACAL RESPIRATION A strategy used by some turtles for absorbing oxygen underwater, in which water is taken in through the cloaca. The water enters two sacs lined with finger-like projections which move oxygen into the bloodstream.

CLUTCH A group of eggs laid in a single session by a single turtle.

CUTANEOUS RESPIRATION A strategy used by some turtles of absorbing oxygen underwater, in which oxygen is passed from the water across permeable areas of a turtle's skin into the bloodstream.

ECTOTHERMIC Without the ability to internally maintain a constant body temperature. Ectothermic animals must use their environment to adjust the temperature of their bodies.

ELEPHANTINE Usually referring to legs and feet. Sturdy, upright and column-like, typically without long toes or webbing. Elephantine legs are found in tortoises and some other terrestrial turtles.

ENDEMIC Found only in a particular geographic location.

ENDOTHERMIC Able to internally maintain a constant body temperature, regardless of the environmental temperature.

GENERALIST Regarding diet, an animal that eats a wide variety of different foods.

HERBIVOROUS Eating only plants.

HIBERNATION A period of winter dormancy. True hibernation involves a slowing of all body processes such as breathing and heartbeat.

INVERTEBRATE An animal without a backbone.

LATERAL On the side.

MARGINAL SCUTE On turtles, the outermost scutes on the carapace.

MARINE Primarily living in a saltwater environment.

MELANISM The condition of being entirely black because of the increased expression of the pigment melanin. Melanistic animals can be born with the condition or, in some turtles, can develop it with age.

OMNIVOROUS Eating both meat and plants.

PELAGIC Living in the open sea or ocean.

PLASTRON The lower portion of a turtle's shell.

SCAVENGER An animal which feeds primarily on carrion.

SCUTE Any of the large, hard scales or plates that cover the shells of most turtles.

SERRATED Toothed in appearance, like a saw.

SEXUAL DIMORPHISM A condition where the male and female of a species look different to one another, whether in size, shape, color, or some other characteristic.

SPECIALIST Regarding diet, an animal which is adapted to eating only or mostly a particular type of food.

TERRESTRIAL Primarily living on land.

THERMOREGULATION The act of regulating one's body temperature, whether internally or using the surrounding environment.

VERTEBRATE An animal with a backbone.

INDEX